Poisoned
Legacy

Poisoned Legacy

The Human Cost of BP's Rise to Power

Mike Magner

ST. MARTIN'S PRESS 🎄 NEW YORK

www.stmartins.com

Design by Maura Fadden Rosenthal/Mspace

Library of Congress Cataloging-in-Publication Data

Magner, Mike.
   Poisoned legacy : the human cost of BP's rise to power /
   Mike Magner.
      p.   cm.
   ISBN 978-1-250-00082-8
   1. British Petroleum Company.   2. Petroleum industry and
trade—Moral and ethical aspects.   3. Petroleum refineries—
Accidents—United States.   4. Petroleum workers—Health
and hygiene—United States.   I. Title.
   HD9571.9.B73M36 2011
   338.7'622182820973—dc22

                                        2010054461

International Edition: June 2011

10 9 8 7 6 5 4 3 2 1

CONTENTS

ACKNOWLEDGMENTS

MUCH OF WHAT IS PUBLICLY known about the inner workings of BP is the result of diligent work by journalists around the globe, especially in places where the company has major operations, such as Alaska, Texas, the Gulf of Mexico region, and the United Kingdom. Thanks to all the reporters and editors at the *Anchorage Daily News*, the *Houston Chronicle*, the *Times-Picayune* in New Orleans, the *Financial Times* in London, and scores of other media outlets, for continuing to dig for the full story in the face of endless corporate spin and occasional obstructionism. Senior BP officials, including former CEO John Browne, declined requests for interviews for this book.

Sincere appreciation goes out to all those who assisted my own reporting efforts along the way, including Ernie Lowe, who shared the research for a paper on BP that he and a colleague published in 1998; Paul Todd and Richard Fineberg, two of the most plugged-in investigative reporters in Alaska; Martha McCoy of Chanute, Kansas, for assisting with coverage of the *Neodesha v. BP* trial; Patti Bonnett, the epitome of a conscientious public official in the clerk's office at the Kansas Judicial Center in Topeka; and Alexandra Proserpio Fox, librarian extraordinaire at the Arlington Public Library in Virginia.

This book would never have been written if not for the relentless pursuit of justice by Lucille Campbell of Neodesha, Kansas, one of the most courageous people I have met in more than thirty-five years as a journalist.

Nor would it have come to fruition if literary agent Ron Goldfarb hadn't recognized its potential and provided unwavering support after I first presented him with the story of Neodesha and BP in 2007.

Finally, special thanks to Denise, Sean, and Emma Rose for their incredible patience, love, and understanding.

INTRODUCTION

ON A GORGEOUS FALL DAY in 2010, fisherman Johnny Nunez sat in a rocking chair on his front porch in Violet, Louisiana, on the Mississippi River delta just above the Gulf of Mexico. Normally under such pure blue skies, Nunez would have been doing what he had done for most of his fifty-five years, plying the ocean or its bays and estuaries in search of speckled trout, redfish, and other prize catches. But six months earlier, a BP oil well blew out in the deep waters of the Gulf about forty miles off the Louisiana coast, and one of the world's most productive fisheries had been turned into a disaster area, poisoned by more than 200 million gallons of heavy crude and untold volumes of natural gas that erupted from below the ocean floor.

Although Fishing Magician Charters run by Nunez and his two sons was grounded by the spill, Nunez himself was one of the luckier victims of the calamity—he was hired by BP for cleanup work that earned him a solid paycheck during the spring and summer. Many thousands of others in the Gulf region were far less fortunate.

Eleven men died almost immediately in the horrific explosion that the blowout caused on the *Deepwater Horizon,* the giant rig that BP was leasing to drill its well a mile below the surface of the ocean; eleven families, including a number of very young children, had their lives irrevocably altered by the loss of their loved ones.

A fishing industry that provides livelihoods for thousands of commercial fishermen, charter boat operators, and seafood processors was crippled, with a corresponding impact on restaurants and other businesses reliant on Gulf shrimp, oysters, and ocean fish. Travel destinations such as the famous white-sand beaches of Alabama were deserted for the peak season, severely crimping revenues at hotels, recreation sites, and companies catering to visitors. There was a domino effect on scores of other Gulf Coast businesses, from health clubs to wedding planners, still reeling from a brutal recession and record unemployment.

All of this social and economic havoc that rippled across the country was an especially bitter pill to swallow because it was caused by a multinational corporation that had promised so much more. Only a decade before the catastrophic Gulf spill, British Petroleum had rebranded itself as simply BP, with a sunny yellow-and-green logo and an ad campaign that pledged to move the world "Beyond Petroleum" into a cleaner energy future. It was one of the first and most successful green-marketing efforts in the short history of the environmental movement, and it had followed a stunning break by BP from the rest of Big Oil when its maverick CEO, the future Lord John Browne of Madingley, announced in 1997 that the company would take a lead role in tackling global climate change.

Certainly BP was not the first major corporation in the new millennium that failed to live up to its carefully crafted public image.

Enron Corporation went from being one of *Fortune* magazine's "most admired companies" to being accused of one of the biggest financial frauds in history in 2001, and eventually the energy-trading giant went bankrupt and settled with investors for $7.2 billion. Overleveraging of high-risk investments by some of Wall Street's most trusted firms led to the bankruptcy of Lehman Brothers, a taxpayer bailout of the American International Group (AIG), and fraud charges against Goldman Sachs following the collapse of the financial industry in 2008.

BP's sins in the Gulf of Mexico were more egregious, though, because twice in the previous five years the company had caused catastrophic accidents by cutting costs to boost profits at the expense of worker safety and environmental stewardship. A March 2005 explosion at BP's refinery in Texas City, Texas, that killed 15 workers and injured 180 others was blamed on negligence of safety standards, and resulted in billions of dollars in fines and legal settlements against the company. A year later, a leak from a BP pipeline in Alaska caused the largest oil spill on land in the state's history and was found to be the result of the company's failure to properly maintain equipment in the harsh Arctic environment.

After both of those disasters, BP apologized profusely to the public and vowed to upgrade safety and environmental programs at all of its operations in the United States and around the world. Following the Gulf spill, as evidence mounted that the company may have been cutting corners to save money and speed completion of its deepwater well when the blowout occurred, the ritual played out again: public apologies, denials of negligence, and promises to "make this right" with cleanup of the environment and compensation for victims.

By October 2010, three months after the leak in the Macondo well field had at last been plugged, BP was winding down its efforts to skim oil from the ocean surface and remove tar balls and petroleum gunk from the beaches and shorelines. The government had declared that seafood from the Gulf was safe to eat, and President Barack Obama and his daughter Sasha had gone for a swim at Alligator Point near Panama City in the Florida Panhandle.

Johnny Nunez was buying none of it. After spending the summer as a BP cleanup contractor and feeling the sting of the chemical dispersants that had been poured into the water to help break down the oil, Nunez said he would not eat fish from the Gulf, and that many of his fellow fishermen felt the same way. "They tell you, don't get in the water, don't touch the water without gloves, and now they're saying it's okay to fish? It's a contradiction," he said. BP was paying for independent lab tests on Gulf seafood, and federal scientists were insisting there was no evidence of harmful contaminants, but no matter. Several months of seat-of-the-pants, trial-and-error efforts to stop the leak meant that both the company and the government had lost credibility during the Gulf crisis of 2010, making it difficult for Nunez—and for many Americans—to take anything that either said for granted.

Heading into 2011, the oil industry and the regulators overseeing it were scrambling to regain the public's trust and confidence that future energy needs could be met without destroying the planet. For BP, the key question was whether it would finally learn from the grave mistakes of its past and truly put the promises and rhetoric of its PR spin into action.

Poisoned Legacy

chapter one

THE BOOK OF COMMON PRAYER

THE OSAGE INDIANS, HAVING BEEN pushed westward by the Louisiana Purchase and a flood of white settlers in the early 1800s, landed on the lush plains of southeastern Kansas between two rivers that came to be called the Verdigris and the Fall. The fertile region had bountiful wildlife and abundant fresh water, both above and below the ground, so the tribe found it a fair substitute for its native lands nearby in Missouri. The verdant geography also made the so-called Indian Territory attractive to the Europeans. After the Civil War, enough whites had migrated to the area that a trading post was built at the convergence of the two rivers, near the base of a hill where the Osage chief known as Little Bear was buried.

The Osage tolerated the whites, but were appalled by their blatant disrespect for the land and waters. The settlers treated the rivers like sewers, dumping their wastes in the water with complete disregard for others who might be living downstream. So when the whites approached the tribe with a somewhat patronizing request for an Indian name for the new town, the Osage leaders decided to make a statement. "Nee-Oh-Da-Shay," the tribe suggested, saying the name meant "Flower of the River." A more precise translation would be "Poop in the River," according to an Osage chief in the 1950s, Freddie Lookout. Most of the whites never caught on, and over time the official definition of Neodesha became "meeting of the waters," which is how the community describes it today.

The community grew quickly, with the two populations coexisting separately but peacefully, at least until the U.S. government decided the Osage Nation should be forced out of Kansas—partly in retaliation for one band, the Greater Osage, having fought for the Confederates during the Civil War—and the tribe was ordered to relocate to a reservation in Oklahoma. Many of the Osage were ready to move anyway after an incident in the early 1870s in which the whites desecrated the mound where Chief Little Bear had been buried in 1868. Members of the tribe had returned for an annual ceremony and found the burial site had been opened and the chief's body stolen, along with other items that had been placed in his grave for his journey to the next world. The Indians were so furious they did a war dance and prepared to burn down the town. It was only after a doctor produced the chief's skeleton that the Osage backed down, but they remained bitter and cursed the settlement for decades to come.

The town's real future was determined in 1891 when drillers seeking a natural gas supply for Neodesha struck oil on land owned by the local blacksmith, T. J. Norman. The site, named Norman No. 1, became the first well west of the Mississippi River to produce commercial quantities of oil—about twelve barrels a day to start—and set off a boom that had Kansas producing 4 million barrels a year by the turn of the century. In 1897, Standard Oil Company built a refinery in Neodesha, making the town one of the first major engines of the industrial revolution rapidly sweeping westward across the continent.

Oil refining has always been one of the messiest processes in modern manufacturing, and it was especially so in its early years when crude oil was mainly converted into kerosene, while other products, such as gasoline, were considered wastes that were typically dumped onto the ground or into the nearest river. Once the automobile industry emerged in the early 1900s, refineries shifted production to gasoline, but plenty of chemical by-products, including toxic compounds such as benzene, still poured into the air and water.

The Neodesha refinery kept getting bigger and dirtier as the decades flew past, and the town grew along with it, turning into a bustling city of three thousand by 1920. Two railroads crisscrossed the community, along with pipelines built to carry fuels to markets throughout the Midwest and beyond. New sections were continually added to the refinery, turning it into a complex of buildings, tanks, pipes, and equipment spread over two hundred acres. The plant supplied special fuels for the U.S. military during the two world wars and gasoline, diesel, and other products to the Farm Belt for more than seventy years. All the while, foul-smelling air, greasy water, and dead or damaged animals became commonplace. John Manderscheid, who grew up in Neodesha in the 1960s, recalled that he and a friend used to play in the creeks near the refinery amid dead fish and deformed frogs. Manderscheid was later diagnosed with spastic paraplegia and moved to Arizona after his doctors recommended a drier climate; his friend died from a brain tumor.

Some made the most of it: the Pettit family on Osage Street, only a block from the refinery, was able to pump gasoline directly from a cistern on their property and use it in their Model T, Robert Pettit said in a letter to the Neodesha newspaper. A photo of his cousin, Russell Grokett, pumping fuel from the Pettit's backyard was published in *Oil & Gas Journal* in the early 1930s, he said. Things that would seem abnormal or hazardous in most communities were accepted in Neodesha as the price of progress. Those who wanted them had good jobs, and most people believed they were living the American dream. Little did they know that for some of their descendants, it would become a nightmare.

In the 1960s a revolt against industrial negligence spread across the land, as people grew increasingly concerned about air thick with smog, rivers catching fire, wildlife disappearing, and a frightening rise in deadly cancers and once-rare diseases. In response, Congress passed some of the toughest environmental laws in history—the National Environmental Policy Act in 1969, the Clean Air Act in 1970, the Federal Water Pollution Control Amendments in 1972 (expanded to the Clean Water Act in 1977). The national crackdown on pollution hit industries like a hurricane, forcing factory cleanups and, in some cases, plant closures. That's what happened in Neodesha in 1970 as Amoco, a midwestern descendant of the Standard Oil breakup in 1911, just walked away from its aging, leaky refinery.

In September 1969 the Amoco manager, William A. Burns, announced the refinery would be closing the following April. He cited economics as the sole reason—a shift of markets to urban areas, higher production costs, less efficiency at the plant compared to larger new ones, declining oil supplies in the region. Burns pledged to help all workers find new jobs, and about two hundred asked for help. Many would move to Texas refineries, some to Missouri, North Dakota, and other states where Amoco had operations, and some managers were transferred to the company's headquarters in Chicago. Nothing was said about pollution, at least not publicly. Amoco did donate $5,000 for razing old buildings on the site, in hopes that it could be turned into an industrial park. Eventually the city's development commission would attract a fiberglass boat company, now Cobalt Boats, along with an aerosol can company and a cabinetmaker. Amoco estimated the appraised value of the site at $1 million, and deeded it to the city, more than four hundred acres in all.

This was a pattern that was repeated in dozens of communities across the United States over the next three decades. Partly because of stricter environmental regulations and partly due to consolidation in the oil industry, the number of active refineries dropped from 276 in 1970 to 158 in 2000, leaving more than a hundred abandoned sites. These heavily contaminated properties dot the landscape from Bakersfield, California, to Marcus Hook, Pennsylvania, and from Alma, Michigan, down to Lake Charles, Louisiana, and Fort Worth, Texas.

In some cases modest cleanups were done, when environmental agencies decided that oily sediments in streams or toxic contaminants in groundwater could pose threats to public health. In rare instances, more extensive restoration was done, such as when a Wyoming

judge who was fond of fishing in the North Platte River ordered
Amoco to spend millions removing all the pollution from the river
and from groundwater near its former refinery in Casper.

But in most cases, abandoned refinery sites were like Neodesha's,
where very little cleanup was done. And at many sites, as in Neodesha, new industries and businesses and even schools and homes
were built on or near the contaminated properties, and the new ten-
ants had no idea they were sitting on a toxic time bomb. "What you
can't see won't hurt you" was the prevailing attitude in Neodesha.
State health officials did require Amoco to monitor the old refinery
site, and at one time it was suggested that a federal cleanup could
be done if the site was added to the Superfund list, but city fathers
wanted nothing to do with the idea, for fear Neodesha would be
branded as a hazardous waste dump like Love Canal, New York, or
Times Beach, Missouri.

And so it went—out of sight, out of mind—for almost thirty
years after the refinery closed, or rather for a full century from the
time the refinery was built in 1897. It wasn't until 1999, a year after
British Petroleum acquired Amoco in the largest industrial merger
in history at the time, when the issue of contamination from the
Neodesha refinery truly surfaced. The new BP-Amoco was now the
second-largest oil company operating in North America behind
ExxonMobil, also the product of a late-'90s megamerger. The power
of these conglomerates made them intimidating to the average consumer, promoting a kind of grudging acceptance of Big Oil. Everyone needed gasoline, after all, so the downsides—price spikes, oil
spills, global warming, refinery pollution—were viewed by many as
the costs of doing business. Savvy BP officials in London, though,

recognized that many Americans might be resentful about a foreign company controlling so much of their lifeblood. So this is when they launched a public relations campaign to give the company a softer, greener image. Ads suggesting that BP stood for Beyond Petroleum—with the British part of the name never mentioned—appeared on television and billboards and in magazines and newspapers. The company adopted a green and yellow sun logo with the friendlier lowercase letters *bp* at the top.

One of the corporation's first steps after the merger was performing "due diligence" and evaluating the assets and liabilities it had acquired from Amoco. BP dispatched a team of appraisers and engineers to Neodesha to determine what was there, and based on what was revealed by later studies, they must have been stunned by what they found. Levels of contaminants such as benzene, toluene, and arsenic were thousands of times higher than the amounts considered safe by the U.S. Environmental Protection Agency, according to studies done later by health officials and the city's engineers. The plume of pollution ran beneath at least 70 percent of the city, with the strong possibility of other toxic "hot spots" in the surrounding area as a result of past dumping by the refinery operation. Most homes and businesses in Neodesha were served by a municipal water system that drew from the Verdigris River, but many older properties still had abandoned wells that could overflow after storms or, worse, serve as conduits for poisonous gases. There was also a good chance that water leaking into basements was highly contaminated. Often after rainy weather, when the groundwater table rose, oily liquids oozed into the cracks in foundations of homes and businesses near the site. At the home of John W. Smith, for instance, two blocks from the old refinery, petrochemicals leached into the basement, dripping in oily liquids from the walls and permeating the house with chemical smells. Smith had prostate cancer and a tumor under his ear and his wife, Sarah, seventy-five, constantly had trouble breathing, so the couple decided to pull up roots and move to Texas for their final years.

BP-Amoco took steps to show it was addressing the problems, and worked with the city to obtain permits and authority for some preliminary cleanup work. It hired a contractor to begin pumping out contaminated groundwater, and received permission from the city and state to treat it on site and release the wastewater into the river. Tons of chemicals that were removed had to be trucked to a hazardous waste landfill. However, the company told the local paper for a September 23, 1999, story that most of the activity was "routine monitoring" of the site. Some groundwater remediation was being done, but there was "no health or safety concern to Neodesha residents," company spokesman Dick Brewster told the *Neodesha Derrick*. "Amoco will keep the community informed of future monitoring activities."

The company may have assumed that this would be the end of the matter, as the oil industry had grown accustomed to successfully glossing over problems in cities and towns everywhere, with little resistance or concern. Then again, British Petroleum had never encountered anyone like Katherine Lucille Campbell of Neodesha.

A retired schoolteacher with a true-blue Kansan independent streak, Lucille Campbell, known by her middle name, read the story in the *Derrick* about BP's "monitoring" and immediately had questions. How could they be so sure there were no health or safety concerns? Lucille had seen the warnings posted on most gasoline pumps saying that benzene vapors can cause cancer and other illnesses. If this was just routine monitoring, why were they removing contaminants?

The questions gnawed at her for weeks, and no new information was published in the *Derrick* as BP-Amoco had promised in the September 1999 story. Lucille was not by nature a rabble-rouser or an organizer; she had never formed a group of any kind or campaigned for any cause, other than sending a few checks to the Sierra Club after joining the environmental group in her fifties. As she entered her sixties, her spinal cord began deteriorating in a painful way that hospitalized her several times and ended up leaving her in a wheelchair much of the time by age sixty-five, making her an even less-likely candidate to become a community activist. She only became interested in the pollution after reading about it in the local paper, and she grew more concerned only because BP's statements that the pollution posed no risks to residents just didn't seem plausible. Having such deadly chemicals so close to the surface, especially in areas where many children were exposed, had to be of some concern, she reasoned. And when she heard that residents near a similar abandoned refinery outside Kansas City had successfully sued BP for property damages, she realized that "maybe my ideas weren't all that off base."

After teaching for twenty years in six different Kansas schools, Lucille Campbell knew how to do her homework. She began talking with people in the community and researching old issues of the

Derrick at the library, and kept coming across stories about people—especially young people—with debilitating diseases such as cancer, diabetes, and multiple sclerosis.

There was Susan Baker, who moved to Neodesha when she was four years old in 1975 and developed diabetes nine years later. Her baby boy Marcus had to undergo open-heart surgery and a liver transplant as a result of birth defects and a disease called biliary atresia; her husband died of cancer at age twenty-four in 1993.

There was the family of Gerry Claiborne, whose oldest son was diagnosed with Hodgkin's disease and lymphoma when he was in the ninth grade. After weeks of chemotherapy he began to recuperate and went on to earn a degree from Kansas University, but he was condemned to a life of thyroid medicine, limited weight gain, and soft bones, his mother said.

At least two other young children in town were found to have leukemia in the early 2000s, and both underwent months of chemo before the cancer went into remission. A five-year-old boy who had died in 1993, Stevie Lee, wasn't so fortunate. Lucille found a newspaper story about Stevie that reported he fell and hit his head at the Jiffy Mart, and the doctor who examined him for a concussion discovered that Stevie had a brain tumor. The boy died shortly afterward.

On the farm next to the Campbell home just outside Neodesha, Don Shafer told Lucille that he was worried his family was exposed to toxic chemicals when he was growing up in a house just two blocks from the refinery. "We played in the ponds and never found any frogs or toads," Shafer said. Some years later, Shafer's daughter was diagnosed with an inoperable brain tumor at age thirty-five.

Radiation treatments have shrunk the tumor some, but she regularly travels to Texas for treatments aimed at stopping the seizures she still experiences frequently. Doctors can't say what caused the tumor, but Shafer said he believes the pollution in Neodesha had a lot to do with it.

Shafer introduced Lucille to Rick Johnson, operator of a salvage yard in Neodesha that frequently put him on the site of the former refinery to recover materials from the industries that had located there. Johnson also took a lot of scrap from the city, including old pipes that had been dug up on or near the refinery site and often set off his Geiger counter with unsafe measurements of radioactivity. Johnson began to suspect that many underground pipes in Neodesha were radioactive, a common condition at refineries where oil can leave "hot" residues behind.

Johnson is no ordinary scrap yard operator. An avid researcher who speaks several languages, he made it his business to find out everything he could about the refinery operations and the mess that was left behind. He discovered pools of oily water in ditches and ponds around the site, and digging just a few feet below the surface often uncovered globs of oil or the unmistakable smell of gasoline or diesel fuel. A huge containment pond that had been built by Standard Oil in the early part of the last century was largely devoid of life. And all around the site where some crumbling buildings remained from the old refinery, basement windows were sealed to hide the black oily water that had seeped inside.

For his efforts, Johnson was rewarded with numerous health problems, including diabetes, several heart attacks, and a bout with leukemia. But he persisted in poking around the contaminated site, monitored the comings and goings of officials from BP and government agencies, and reported all that he learned to Lucille Campbell.

Lucille handled the library research, including scrolling through the local newspaper archives, and she compiled a list of deaths and serious illnesses in the town that quickly grew into the dozens. It seemed like a high rate for a community of less than three thousand people, and the problems seemed to be clustered in the area around the refinery site, in neighborhoods filled with homes built decades ago for the plant's workers and their families.

Most disturbing of all to Lucille was the discovery that her own three-month-old daughter, who died unexpectedly when the family lived a block from the refinery in 1964, may have been a victim of the pollution.

FUEL AND CHEMICAL SMELLS WERE thick in the air over Neodesha when Gretta Campbell was born on November 8, 1963, much as they had been almost since the turn of the century when Standard Oil built the refinery. Her mother, Lucille, didn't think of it then, of course, but looking back it was no wonder that Gretta didn't start breathing properly when she arrived, and the doctor had to force her to use her lungs. Lucille Campbell's only concern at the time was that her fifth child in five years would be as healthy as the others. Sadly, she would learn all too soon that would not be the case.

Gretta was a smiling baby, to be sure, but one who was strangely quiet. She rarely fussed or cried, even when hungry or wet. Lucille wondered if something was wrong—although the doctor hadn't said anything was—and worried about it constantly at first. Had she caused a problem during the pregnancy? Was she providing the proper care? Was she giving Gretta as much attention as she gave four-year-old Robert, three-year-old Janice, two-year-old Bill, and one-year-old Tommy (or was that even possible)? The questions were always in the back of her mind, although the concerns gradually subsided as Lucille was swept up by the job of caring for five young children. Then, two weeks after Gretta's birth, President John F. Kennedy was assassinated, and everyone was overwhelmed by the pall cast over the entire country. Nothing else so horrible could happen, Lucille thought.

As winter settled over the plains, Lucille and her family became decidedly unsettled. Bob Campbell, a gandy dancer for the Frisco Railroad, was being transferred to Wichita for construction work on the rails. Even though Neodesha was her hometown, Lucille was looking forward to leaving. It wasn't just the constant odor, so oppressive at times that you couldn't go outside on an otherwise gorgeous day. The refinery itself, only a block away from the Campbell home, was an imposing, threatening presence she would be glad to have in her past. Lucille had many memories of the huge plant operated by Standard Oil and one of its successors, Amoco, since 1897. And most of them were bad. There was the time when she was six years old, in the middle of World War II, when she was waiting for her parents near the refinery site and was stomping around on the cover of a well. The cover collapsed and down she went, right into an oily, smelly brine. Fortunately, her sister saw Lucille's bright red hat as it was disappearing into the ground, and she was pulled out quickly, but not before swallowing some of the nastiest liquid she would ever taste. It was one of her earliest and worst memories. Maybe if she moved away from Neodesha it would sink deeper into her subconscious.

By February the family had moved many of their belongings to Wichita, and on Friday the twenty-first they came back to clean out more of the four-room house they had lived in since 1961, in a neighborhood near the refinery called "Glasstown" because a glass factory operated there years ago. Once the oldest four had settled down to sleep in one bedroom, Lucille took the baby into her and Bob's room and laid Gretta on a platform rocker that once belonged to Lucille's grandmother, because they already had moved the baby bed to Wichita. The rocker had large, padded arms that extended into the seat, and Lucille had used it many times as a child to curl up and read or take a nap. She put the chair with Gretta next to the bed when she and Bob retired.

Lucille awoke Saturday morning and found Gretta asleep. "Hey, sleepyhead," she said as she picked the baby up. "I can't believe you aren't fussing for a bottle or dry britches!" But Gretta felt cold and did not respond. Lucille knew immediately that something was wrong. She wrapped the baby in her robe and kept trying to wake her. Gretta was breathing but her body was limp. Lucille told Bob to stay with the other children while she drove Gretta to the hospital. Dr. Charles Stevenson, who had delivered Gretta three and a half months earlier, examined and admitted her, but started no medical procedures. She was placed in a semiprivate room with a young woman who had attempted to commit suicide.

Lucille waited with Gretta for what seemed an eternity. Her color began to turn blue; her breathing became shallower. Lucille kept talking to Gretta and trying to wake her. The only noticeable response was a few tears that ran down the side of her face. Suddenly she appeared to have stopped breathing. Lucille ran into the hallway and cried for help. Dr. Stevenson had left the hospital, so Dr. Frank Moorhead came and called for a crash cart. Dr. Moorhead worked on Gretta, tried to force her to breathe as Dr. Stevenson had done at her birth, but it was futile. He later told Lucille that it appeared Gretta had "bled out" in the brain, much like an older person having a stroke. He called it a congenital aneurysm. No autopsy was done, however, and Lucille never asked for one because she couldn't stomach the idea of her beautiful little girl being dissected. No one mentioned it to her then, but Dr. Stevenson signed the death certificate and listed the cause of death as an accident— that Gretta had caught her head in a chair.

It wasn't until thirty-five years later that Lucille learned what really might have caused Gretta's death. Lucille and Bob had returned to Neodesha in 1987, after the other four kids had grown up. Lucille ended her career as a teacher, and Bob was forced to retire with a disability after being injured on a job with the Kansas highway department. In 1999, Lucille became interested in genealogy, inspired by all the stories she had heard over the years about her roots. Ancestors on her mother's side helped establish the city of Hartford, Connecticut, and fought in the Revolutionary War (making Lucille a member of the Daughters of the American Revolution, Daughters of the American Colonists, and the National Society of New England Women). She traced her father's side back to Brooklyn,

New York. She followed the family's path through Civil War battles, with one great-great-grandfather fighting for the North and a great-grandfather for the South. After the war, some of the family made its way to Kansas by wagon train, and ended up in Neodesha, where Lucille's father worked at the W. J. Small alfalfa mill.

One thing that struck Lucille during her research was that there were no reports of serious illnesses on her family tree. She thought about Gretta again and old doubts that had nagged her for decades came rushing back: Had she done something wrong? Did she provide proper care? She remembered the doctor describing an aneurysm. What could have caused that in a three-month-old baby? Lucille needed to know more, so she wrote to the Wilson County Hospital and asked for records of Gretta's hospitalization and death.

A short time later, the hospital called Lucille to say that the records of Gretta's treatment had been found on microfilm. She went immediately to look at them and found some so faint they were unreadable, but the lab results were crystal clear. "I felt like someone had poured ice water on me," Lucille later wrote in an article about Gretta's death. "Her white [blood cell] count was 44,700!" One of Lucille's past jobs had been in a family doctor's office, so she knew what this meant, although she consulted with several doctors and nurses to confirm her suspicion. "The general opinion was that her death could very likely be due to leukemia such as acute lymphocytic leukemia resulting in thrombocytopenia, acute mylogenous," she wrote. In other words, a rare, fast-progressing, and deadly form of cancer.

A wave of anxiety came over Lucille as she connected the dots between the refinery pollution and Gretta's illness. And as she gathered information about other illnesses and deaths in the town, she realized that Gretta was only a small part of the total picture.

Lucille Campbell researched cancer rates and average ages of death in Kansas cities and counties, and found evidence that Neodesha's and Wilson County's were disproportionately high. She formed a group she called NEAT—the Neodesha Environmental Awareness Team—and wrote up newsletters that she distributed at stores, the library, and city hall. She contacted Erin Brockovich, whose campaign to help a California town sue a utility for toxic dumping was made into a successful movie, for legal advice. (An attorney recommended by Brockovich, Ed Hershewe of Joplin, Missouri, looked into filing a suit over cancer cases in the town, but decided it would be difficult to prove a clear link to the pollution.) Lucille wrote the Kansas Department of Health and Environment, the U.S. Environmental Protection Agency, and every other government agency she could find, asking for an investigation.

What followed was one of the most remarkable challenges to the power of the oil industry ever seen in the United States, one in which a British company was being asked, in essence, to pay for a century of sins by Standard Oil and its successors in the center of America's heartland. It was a classic David versus Goliath story: A small Kansas town, prodded by a senior citizen in a wheelchair, files a $1-billion lawsuit against a giant international conglomerate. And the drama of *Neodesha v. British Petroleum* would be played out by a cast of characters worthy of Hollywood, with Lucille Campbell perfectly suited for the leading role. She was the quintessential Great Plains woman, a descendant of Revolutionary War and Civil War heroes, whose grandparents came to Kansas in a covered wagon, put down roots, and thanked God for delivering them to the Promised Land. And when the promise turned into a lie, she rose up defiantly and demanded justice.

BP, of course, took none of this lightly. The company moved quickly to try to show it was addressing any potential problems. It presented a cleanup plan to the city commission calling for air testing in homes and businesses and groundwater pumping to remove contaminants. While constantly assuring residents there were no threats to their health and safety, BP quietly arranged for its attorneys and the Neodesha mayor, J. D. Cox, to sign a covenant on April 17, 2000, stipulating that for the next seventy-five years, until January 1, 2075, the refinery site would not be used for residences, farming, or overnight lodging; no structures would be built with basements; only limited excavations would be allowed; and the groundwater would not be used. The company described the covenant as merely precautionary, but there was a larger purpose: Restrictions on development at the site would ease pressure from the state for a full cleanup of the groundwater.

BP appeared to believe that if it was going to stop a lawsuit it had to convince the community that litigation was unwarranted and would be damaging to Neodesha's reputation. To help make its case, the company hired a top law firm, Sidley Austin, and the Kansas City office of a national public relations firm, Fleishman-Hillard. A key product of their consultations was a strategy memo written by attorney Evan Westerfield titled "The Book of Common Prayer," which appeared aimed at appealing to the community's Christian faith to squash talk of a lawsuit against BP. Wouldn't it be wrong, they would argue, to go to court against the very industry that built the town in the last century? Westerfield was a partner in the Chicago office of Sidley Austin, one of the nation's oldest and most prestigious law firms, and he was a Yale University and Chicago Law School graduate who specialized in defending corporations charged with environmental crimes. In this case, though, Westerfield's goal was to prevent a lawsuit in the first place, and that meant finding a way to undercut the efforts of Lucille Campbell.

Lucille was convinced that BP's team was closely tracking her activities, looking for vulnerabilities it could attack and information it could use to gain the upper hand when she challenged the company. At the Neodesha library, where Lucille often did research, she began to feel employees were spying on her: Once when she returned to retrieve something, she found a library aide sitting at the computer she had been using, and it appeared to Lucille that the aide was trying to go back over her research. Fleishman-Hillard consultant Julie Gibson followed every letter and column Lucille Campbell wrote to the *Neodesha Derrick* and immediately prepared responses for BP officials to submit to the paper.

Armed with information about Lucille, BP's strategists waited for a chance to pounce. An opportunity presented itself when she was invited to attend a meeting in Sugar Creek, the Kansas City suburb where Amoco had closed another refinery in 1982. Residents near the site had already sued the company for property damages and won substantial settlements, and more than thirty of them were taking it a step further and seeking compensation for health problems they claimed were caused by contamination from the refinery, which had operated for nearly eighty years. Lucille went to a meeting there organized by a neighborhood group called CLEANUP, and had her first real confrontation with officials from British Petroleum.

As soon as she signed in for the meeting, a confident, well-cropped man in his forties hustled up to her and introduced himself as Ron Rybarczyk, director of public relations for BP-Amoco. After a bit of chitchat, Rybarczyk asked Lucille to meet with him and two other BP officials—press spokesman Dick Brewster and Neodesha project manager Lloyd Dunlap—at the Neodesha library on an upcoming Saturday in early 2002. Before Lucille had a chance to respond, some women from the CLEANUP group interrupted, Rybarczyk walked away, and the women warned Lucille to beware of him. Their words of caution only reinforced what Lucille already felt in her gut about Rybarczyk and BP. Lucille had a bad feeling that the company was trying to gang up on her, which made her very nervous.

It turned out her instincts were correct. In a February 2002 memo filed in court years later, public relations experts from Fleishman-Hillard reported to BP that they had spent more than thirty hours in the first weeks of the year developing arguments and strategies to counter Lucille Campbell. Even without knowing this, Lucille sensed that BP was laying a trap for her, and she decided not to go to the library on the designated Saturday. A few days later she received a note from Rybarczyk, saying he "very much" regretted that she hadn't come to the meeting. "I understand you were told not to meet with us all by yourself," he wrote, although Lucille wasn't quite sure how he knew this—she had never mentioned her concerns to him. Rybarczyk offered to let Lucille bring anyone she wanted to a meeting in the near future, so they could "start direct communication with each other, because without that, I'm not sure how anything productive can be accomplished." Lucille did not respond.

BP did not give up easily though. They seemed confident, in fact, that they could win over Lucille through the Bible Baptist Church, where she had been a member for many years, and where her son Bill was training to become a minister. The company felt it had an ace in the hole at the church, for the pastor and Bill's teacher, DeWayne Prosser, was a former city commissioner and a current county commissioner who adamantly opposed a lawsuit against the oil company. "Some of us still are receiving benefits from BP because of years of working for Standard Oil and we think that BP is still taking very good care of their employees," Prosser and a group of residents wrote in one letter to the local paper. "We think that you are 'biting the hand that's feeding you.'"

Prosser's father had been a refinery employee, and many former workers at the plant and their families were members of his church. Prosser himself suffered from Marfan syndrome, a genetic disorder that put him in the hospital on life support nearly two dozen different times, but he was adamant that the disease was not connected to contamination in Neodesha.

Prosser's wife, Elaine, was also very vocal in supporting the company. Once after church, Lucille Campbell showed her a photo of a two-year-old boy who was bald from chemotherapy, and asked if she thought the refinery pollution was killing the town's children with cancer. Lucille recalled that the preacher's wife angrily pushed the picture aside and exclaimed: "Maybe it's God's judgment upon them!"

Not long after that, five-year-old Charles Catron became seriously ill in Neodesha, where he was being raised by his grandparents, Connie and Dennis Catron. After coming down with what appeared to be a cold or flu on Valentine's Day in 2003, the boy grew more and more listless, constantly falling asleep, and unable to eat much solid food. After more than a week in which he showed no improvement, Connie decided he should be taken to the hospital, but a raging snowstorm made her and Dennis hesitant to drive themselves. Connie called for an ambulance, and while they were waiting for it to arrive Dennis was holding Charles in his arms when he stopped breathing.

The ambulance attendants transported the boy's body to the hospital for an autopsy. "They found a 55,000 white blood cell count," Connie Catron said. "It shut his whole system down." She and her husband don't know if Charles's illness was caused by exposure to toxic pollution, but the question haunts them to this day. "His mother was raised in Neodesha," Connie said. "She might have been affected by the pollution. I really don't know. We don't have any answers. I was raised there, too. My mother was raised right there near the refinery and she got breast cancer. They need to be held accountable if that's what caused it."

Meanwhile, BP's PR team decided to appeal directly to Lucille Campbell's Christian faith in its efforts to get her to back off from talking up a lawsuit. This time BP's project manager in Neodesha, Lloyd Dunlap, was dispatched to approach her after a meeting. "Lucille, I came to know the Lord when I was seventeen," he said. It seemed awkward and totally out of place, but Lucille played along, asking Dunlap what church he attended now. He seemed to fumble for an answer, finally saying, "Southern Baptist." Then he flashed a ring with a Christian fish emblem. "I felt like someone had been checking into my private life," Lucille said later.

A few days later Lucille received a note from Dunlap: "I would like to meet with you to explain my walk with the Lord," he wrote. "I would suggest that your pastor come along also, just to listen. Let me know if we can do this." Lucille declined the invitation, and would soon face a brutal retaliation. It came at her son's ordination, an event Lucille went to filled with pride and joy, only to have Prosser spoil the occasion by approaching her beforehand and warning her not to bring up the pollution with any member of the congregation. After the ceremony, Lucille's son told her he agreed with the pastor, and said his mother would have to go before the congregation to consider her banishment from the church. Lucille was devastated; the two most important institutions in her life, her church and her family, had turned against her.

Lucille was beginning to get disheartened, which seemed to be exactly what BP intended. Like a shark smelling blood in the water, BP tried to put an end to the matter once and for all. It presented a report to the Neodesha city commission saying its investigation had found that most of the contamination in the groundwater below the site was caused by a lightning strike on an oil tank in 1968. Lucille immediately searched the library files of the *Neodesha Derrick* and found a May 27, 1968, story about the lightning strike; the paper reported that the tank that was hit contained a blending agent called ultra-formate, but none of the petroleum products now found in the groundwater. Her friend Rick Johnson recalled watching the burning tank as a teenager sitting on the porch of his home directly across the street from the refinery. Johnson remembered that fire-fighters were able to siphon much of the liquid from the tank as it was burning, to enable them to put it out sooner, but it still took all night for them to bring the blaze under control.

Lucille was convinced BP was grossly understating the problems, but city officials appeared to be buying the company's explanations and there was little she could do to change their minds.

Then suddenly—almost miraculously, it seemed to Lucille—she received a response from one of the many government agencies she had petitioned for an investigation. The Agency for Toxic Sub-stances and Disease Registry (ATSDR), part of the federal Centers for Disease Control and Prevention (CDC) and the U.S. Department of Health and Human Services, planned to send a team of health specialists to study illnesses and deaths in the community. The team's leader, Shawn Blackshear, was a lieutenant in the Public Health Service, a dedicated public servant who was passionate about his work. On his first trip to Neodesha from his office in Kansas City, he asked Lucille to take him around the town. Blackshear was shocked by what she showed him. Especially disturbing was a job-training facility with a day care center sitting right where some of the refin-ery tanks used to be located. Test wells drilled by BP and the state health department showed that the pollution in the groundwater was concentrated below the site, but had spread beyond it into the surrounding neighborhoods, Blackshear said. "I was just floored at how huge that underground plume was," he recalled years later.

Blackshear and his team of investigators met with residents, re-searched the refinery's history, and conducted tests around town on soil, water, and air. One sample from a well on the high school grounds showed benzene levels of 20,000 parts per billion, more than two thousand times above the so-called safe limit. All the information was sent to CDC headquarters in Atlanta, where health experts would evaluate it and write a report, called a health consultation, for the community. Blackshear wasn't sure what the consultation would say, but he had a gut feeling there were serious problems in the town that needed to be addressed.

But Blackshear also sensed that BP was bullying the town—and possibly state regulators—into accepting that the pollution posed no threat. Once after Lucille Campbell left a meeting hosted by the ATSDR, BP's public relations manager, Rybarczyk, "started mouthing off about her, trying to paint her as a nut or a goofball," Blackshear said. The comments infuriated Blackshear. "She's concerned about her community!" he said he thought to himself as Rybarczyk ranted about Lucille Campbell.

Blackshear also thought it was curious that Lloyd Dunlap, the BP site manager in Neodesha, had previously worked at the Kansas Department of Health and Environment (KDHE) and while he was there, from 1994 to 1997, he was the supervisor of Pam Chaffee, who was now KDHE's manager on the Neodesha site. The reality, too, was that KDHE relied entirely on BP for data about the groundwater contamination, and the ATSDR had to rely heavily on KDHE, Blackshear said. Even so, KDHE officials acted like the federal agency was stepping on their turf, he said. "It was kind of this thing that this is our territory, the feds can't tell us what to do," he said. The overall impression was that KDHE was defensive about BP, and "BP's paying the bill," he said.

No matter how anyone sliced it, though, there was no question the contamination in Neodesha was vast and extensive. In 2008, in a progress report to the city on BP's cleanup efforts, KDHE itself said that the company had extracted and treated 164 million gallons of contaminated groundwater and recovered 78,237 gallons of petroleum products (enough to fill four average-sized swimming pools), yet the cleanup was still far from complete.

Despite BP's insistence that the pollution did not represent a health hazard, once a federal agency came to town and started investigating, the tides of public opinion in Neodesha began shifting against the company. In January 2003 Lucille received a call from Mayor J. D. Cox, who a few years earlier had secretly signed the agreement with BP limiting development on the refinery site for seventy-five years. Cox told Lucille he had been asked to pass along some concerns the city had about her activist group, NEAT, but the mayor confided that he thought she was doing a good job. He didn't want to be seen with her at meetings, but he was now "wary" of BP. He encouraged her to continue.

BP must have picked up the scent. The next month Lucille was invited to join a "corrective action advisory board" set up by BP and the city to seek ideas for how the "final stage" of the cleanup should be conducted. Lucille agreed to participate, but could tell from the start that the effort was heavily skewed toward BP with Dunlap in charge. Pam Chaffee of KDHE, Dunlap's former employee, was also on the panel. The group also included several residents who clearly sided with the company. After several months of meetings, the advisory board rubber-stamped a series of recommendations largely drafted by BP, including suggestions that topsoil contaminated with lead be removed from the site and that deposits of oil found just below the surface in some areas be cleaned out. An attorney for the city would later equate the plans to playing with Tinker Toys on an industrial site.

Later in 2003, the ATSDR gave its health consultation to the city, and it was a bitter disappointment for Lucille and her supporters. The report said there were concerns about cancer-causing chemicals being present throughout the community, and it recommended further monitoring, but the agency said it could not make a clear link between the contaminants and deaths or illnesses in the community. "Because additional surface soil sampling, groundwater and ambient and indoor air sampling are needed to make a public health determination, the former Neodesha refinery is classified as an indeterminant public health hazard based on data limitations," the report concluded.

The leader of the investigative team, Blackshear, was utterly dismayed and frustrated with his agency. "Every time they would go to a site they would say the same thing—no public health impact," he said later. A few months after the Neodesha report was issued, Blackshear quit his high-level post in the agency's regional office and took a job with the Bureau of Indian Affairs in a remote part of Utah.

By early 2004, Neodesha was split into two camps, those for and those against a lawsuit. BP had many prominent citizens on its side, including Terry Harper, the publisher of the *Derrick,* who asked Lucille if she was trying to destroy the city's economic future. Her former pastor, DeWayne Prosser, was also still a strong force on the Wilson County Board of Commissioners, keeping a majority there in line against litigation. Prosser argued that some pollution was inevitable from the messy business of producing fuel, but the industry should be trusted when it said there was little risk. "Until we quit driving vehicles and using plastics, we have to get the oil out of the ground and refine it as safely as possible," he said in one of his campaign statements for another term on the board that year.

But a number of city officials—including J. D. Cox, who switched from mayor to city administrator in 2004, and new mayor Casey Lair—were beginning to feel that they had been misled by BP. A company risk assessment in 2003 had staunchly maintained there was no threat to public health from the pollution, yet by the end of the year federal investigators were expressing concerns, however muted. BP also insisted that nearly all the pollution was caused by a lightning strike on a tank in 1968, but subsequent tests by the city's own engineers revealed so many different chemicals in the groundwater that they could not possibly have been caused by a single spill.

City leaders were especially incensed that BP failed to warn them about contamination that clearly could pose a threat to the community's young people. "It is difficult to understand how BP, which was fully aware of the severity of the contamination under the old high school, was not forthcoming with that information, instead choosing to remain silent while we unknowingly built and expanded our new facility atop some of the most serious contamination," former mayor Cox and current mayor Lair wrote in an open letter to the community in 2004. "We attempted good faith negotiation with BP but shockingly watched as they methodically staffed their side of the negotiating table with paid professional negotiators whose experience was earned working for BP in settling other closed refinery sites."

On the strength of those arguments, the city commission voted in March 2004 to sue the company for "fraudulent concealment, fraud by silence, negligence, nuisance and trespass." The lawsuit demanded a cleanup and the city's attorney, John M. Edgar of Kansas City, filed a statement with the court estimating damages to be at least $1 billion. Within months the local school district and several businesses on the former refinery site joined the lawsuit. And after Lucille's former preacher, Prosser, lost his reelection bid to the county commission, the Wilson County board voted to become a plaintiff, too.

The Edgar law firm, made up of John M. Edgar and his two sons, David and John F., was hired by the city because of its experience in class action suits, and because the Edgars agreed to take the case on an incentive basis. "They will be paid on results, not just efforts," City Administrator Cox assured the community. The Edgars' legal team spent the next two years interviewing Neodesha residents and officials, gathering documents, and investigating the contamination with its own engineers. They discovered that the plume of pollution beneath the town was much bigger and much more toxic than BP had acknowledged. In fact, their consultant's studies found that there was a layer of "pure petrochemical sludge" riding above the groundwater, a poisonous plume that was up to two feet thick in several places, and was measured at almost six feet thick at one test well.

BP's attorneys, meanwhile, maneuvered to have the case dismissed or moved to federal court, without success. During the discovery process, the company's legal teams—from Sidley Austin, led by West-erfield; and from Squire, Sanders & Dempsey of Los Angeles, led by former military lawyer Steven A. Lamb—inadvertently turned over the memo outlining BP's strategy, the one that Westerfield had dubbed "The Book of Common Prayer." When the Edgars inquired about it, Westerfield moved that it be sealed in the court file, argu-ing that it was a confidential document covered by attorney-client privilege.

Because the memo was sent to BP's "community relations team," including public relations manager Rybarczyk and the community relations manager for one of BP's contractors, the RETEC Group, David Edgar argued in court that "The Book of Common Prayer" was clearly a public relations strategy. "The toothpaste is out of the tube," Edgar argued before District Judge Daniel D. Creitz, who was presiding over the case. "There's been an extensive disclosure." Edgar added, "And I think here this document, as I said, because of its content is shocking. And I don't think that's the kind of docu-ment that fairness would require the return of. I think it's evidence of the fraud we have alleged."

Creitz ended up granting BP's request for the document to be sealed, but the cat was literally out of the bag. In the Edgars' view, the memo proved that BP had perpetrated a fraud.

While the legal work went on, the lawsuit wasn't in the local news very much, but it was still on many people's minds. Some were hopeful for a cleanup; others were angry and resentful toward Lu-cille Campbell. After she wrote a newsletter about her baby Gretta, saying the lawsuit might finally bring some justice for Gretta and other children in Neodesha who had died or contracted serious illnesses, Lucille received an anonymous note in July 2006 that pierced her heart like a knife. The typed letter read:

WHEN YOUR BABY DAUGHTER IS EXHUMED FOR EXAMINA-
TION, IT WILL BE FOUND THAT LEUKEMIA DID NOT KILL HER
BUT YOU DID THE EVIL ACT BY YOUR OWN HAND.

Lucille was crushed. Was this an attempt at intimidation by BP, she wondered, to try to give her a taste of what was to come if the case went to trial? She broke down and sobbed after reading the letter, then became more determined than ever. "Gretta would be forty-three now but will always be a baby in my heart," she wrote in a letter to a local church newspaper a few weeks later. "I have cried more than once over babies' and children's obituaries I see in the paper. There are far too many of them. The major reason I fight the deadly pollution is because of the children. It appears the truths I'm speaking out about are getting too close to home."

Two months after the anonymous letter was sent, Judge Creitz, following a hearing in his courtroom in Iola, about forty miles north of Neodesha, certified the lawsuit against BP as a class action, making all owners of property above the contaminated plume eligible to participate as plaintiffs. The trial was set for August 2007.

chapter two

THE RISE OF BRITISH PETROLEUM

THE BRITISH ONCE DEFINED IMPERIALISM, so it follows that their biggest corporation would have grown and thrived over the past century largely by exploiting the resources of other nations.

The seeds of the company now known as BP were planted in Persia 110 years ago by English capitalist William Knox D'Arcy. Fresh off a successful gold-mining adventure in Australia, D'Arcy had learned through his connections in London that the shah of Persia, desperate for an economic turnaround, was looking for an investor in the oil, gas, and other minerals believed to be abundant in his desert kingdom.

"The D'Arcy Concession" that the shah signed on May 28, 1901, gave the gentleman exclusive rights to petroleum, natural gas, as-phalt, and ozokerite—a mineral wax used in cosmetics—throughout most of Persia for sixty years. In exchange, D'Arcy would provide the shah with a lump sum of £20,000 as soon as the first extraction company was formed, another £20,000 worth of shares in the company, and 16 percent of future net profits.[1]

The agreement required the first company to be formed within two years, and D'Arcy barely made the deadline by registering the appropriately named First Exploitation Company on May 21, 1903. The delay reflected the difficulties D'Arcy had raising funds for the enterprise, and the struggles would continue for another five years. Problems maintaining the frail drilling technology of those days in extreme climate conditions, combined with the lawlessness that crews faced in remote parts of Persia controlled by warring tribal chiefs, made the search for oil and gas difficult, dangerous, and expensive.

Nearly broke by 1908, D'Arcy was under pressure from his investors to find oil or abandon the effort. Then, a little more than a month after he received such an ultimatum in April, his chief engineer in Persia, George Reynolds, finally struck black gold—his drillers hit a huge oil field at Masjid-i-Suleiman, about 130 miles above the northern tip of the Persian Gulf. Over the next year, work began on a pipeline to Abadan, an island city on the gulf where the world's largest refinery would eventually be built, and the Anglo-Persian Oil Company was incorporated on April 14, 1909, the date BP uses as its official birthday.

After a period of construction prolonged by a manufacturing strike and a cholera outbreak, the refinery was completed in 1912 and Anglo-Persian began shipping crude oil, kerosene, and "motor spirit" to the United Kingdom. Thin sales barely kept the company alive, however, and it was only Winston Churchill's appointment as

First Lord of the Admiralty in 1911 that kept it from bankruptcy. Churchill favored using fuel oil in British ships, and through his efforts the government agreed in 1914 to buy forty million barrels from Anglo-Persian over the next twenty years for £2 million. The British government also became Anglo-Persian's largest shareholder, a position it would hold for the company's next sixty-two years.

World War I sparked more growth, as Anglo-Persian purchased its own fleet of tankers to escape rapidly rising shipping rates. It also acquired a German company that distributed Royal Dutch Shell "motor spirit" in the United Kingdom, but had been seized by the British government as a potential enemy threat. The company, which had about a third of the British petrol market, was called British Petroleum.[2]

After the war, Anglo-Persian started to spread its wings. It formed a partnership with Royal Dutch Shell and the French oil company Compagnie Française des Pétroles, which received a concession to develop oil in Iraq in 1925. Two years later, the partnership, called the Turkish Petroleum Company, discovered a massive field at Kirkuk that would make Iraq a major oil producer, but British and French control over such a large petroleum center made the United States nervous. U.S. officials pressured their friends in the British government, which had majority control at Anglo-Persian, to allow American oil companies to buy half of Anglo-Persian's 46.5 percent share of Turkish Petroleum, giving them a 23.25 percent stake in Iraq's rich resources.

By the early 1930s, Anglo-Persian's German acquisition, British Petroleum, was selling its "BP" brand fuels throughout much of

Europe and moving into Australia and India. A partnership between Anglo-Persian and Gulf Oil also hit a gusher in Kuwait that would spawn the tiny Arab emirate's substantial oil industry.

But while it was becoming a major player on the world market, Anglo-Persian was setting the stage for political turmoil later on its original turf. The company did business on the side with friendly tribal chiefs in Persia, and when word got to the shah, he accused Anglo-Persian officials of denying him the full 16 percent of company profits prescribed by the 1901 agreement. Persian authorities hired a British accountant to audit Anglo-Persian's records and learned that royalty payments were being calculated after depreciation, interest, and British taxes were deducted, and therefore the shah's government was being significantly underpaid. News of the apparent thievery fueled anti-British sentiment in the Persian capital, Tehran, where newspapers began calling for cancellation of the D'Arcy Concession.

Anglo-Persian officials appealed to the British government for help appeasing the shah, who agreed after negotiations in 1933 to a new, sixty-year concession with terms more favorable to Persia. (Two years later, Persia became Iran, and Anglo-Persian became the Anglo-Iranian Oil Company.) The tensions subsided during the buildup to World War II, as the rise of the Third Reich engulfed Europe and the Middle East. The war itself was profitable for Anglo-Iranian, which became the prime supplier of fuels to the British Army, Navy, and Air Force, as well as to some of the Allies. The downside for the company was the strengthening of bonds with the British government, which was increasingly viewed with suspicion and distrust by foreign leaders being pushed toward nationalism, particularly in the Middle East. Animosity toward Britain also rose in the Arab world when the British Mandate to oversee Palestine was ended in 1948 and the Jewish state was established there.

Nowhere was the enmity more evident than in Iran, where public sentiment against Great Britain reached a fever pitch in the late '40s. Anglo-Iranian tried to stem the tide in 1949 by agreeing to a substantial hike in royalties and a large lump sum payment to the Persian government. But the shah, hoping that calls for nationalizing Iran's oil industry might subside, did not submit the agreement to the Parliament's Oil Committee for a year. When he did, Anglo-Iranian's offer was fiercely rebuked and calls for nationalization grew louder.

The United States pressed the company to make a more generous proposal, but Anglo-Iranian's chairman, Sir William Fraser, balked at the suggestion. Fraser was a traditional Englishman who firmly believed that the Iranian upstarts must be quashed—President Harry Truman labeled him "a typical nineteenth-century colonial exploiter." The arrogance just added fuel to the fire that was raging in Iran.[3]

After Iranian Prime Minister Ali Razmara came out publicly against nationalization in March 1951, he was assassinated four days later. The Parliament quickly passed a resolution to nationalize oil operations, and chose the chairman of its Oil Committee, Mohammed Mossadegh, to replace Razmara as prime minister with a mandate to take over the industry. The British Cabinet under Winston Churchill, who returned as prime minister in 1951 after a six-year hiatus, began developing plans for military intervention. At the very least, the cabinet leaders believed, Anglo-Iranian's huge refinery at Abadan must be seized and secured.

President Truman, fearing that a British assault would open the door for the Soviets to move into Iran, begged Churchill to hold off. Truman's secretary of state, Dean Acheson, dispatched his top diplomat, Averell Harriman, to meet with Mossadegh and urge him not to kick out the British. "You do not know how crafty they are," the Iranian prime minister responded. "You do not know how evil they are. You do not know how they sully everything they touch."[4]

Mossadegh was in an impossible spot even if he believed deep inside that he needed Anglo-Iranian to manage the country's oil production. The leader of the Iranian right, Ayatollah Kashani, told Harriman that he suspected Mossadegh might be sympathetic to the British, and if he acted on those urges he would meet the same fate as his predecessor.

The Iranians went ahead and took control of the oil fields, pipelines, and the Abadan refinery, and ordered Anglo-Iranian employees to leave the country in September 1951. The British tempered their response, settling for blockades to prevent oil from being shipped out of Iran. The resulting cutoff of a major oil supply, at a time when the United States was fully engaged in the Korean War, heightened concerns about meeting world energy demands, and Washington's attitudes about intervention in Iran began to change.

When General Dwight Eisenhower succeeded Truman in 1953, Churchill made a new appeal for action, raising the specter of a Soviet takeover of the Middle East. Eisenhower's secretary of state, John Foster Dulles, was on board immediately, arguing that Soviet control of the world's chief source of oil could not be allowed to occur. Eisenhower and Dulles turned to Kermit Roosevelt, grandson of former president Theodore Roosevelt and a top official at the Central Intelligence Agency, to team up with the British intelligence service MI6 and lead Operation Ajax, a plan to assist the shah in ousting Mossadegh as Iran's prime minister.

Roosevelt himself traveled to Iran and snuck onto the Tehran palace grounds under a blanket in the backseat of a black sedan, although he wrote later that the cloak was hardly necessary: "The sentry on duty silently, matter-of-factly, motioned us through."[5] When the sedan stopped on a driveway inside, the shah slipped into the backseat and Roosevelt tossed the blanket on the floor. The shah had never met Roosevelt, but recognized him immediately, making the elaborate arrangements the CIA operative had made to confirm his identity to the shah unnecessary as well. Nevertheless, Roosevelt stuck with the script. He told the shah he was speaking for both President Eisenhower and Prime Minister Churchill. "President Eisenhower will confirm this himself by a phrase in a speech he is about to deliver in San Francisco—actually within the next twenty-four hours," Roosevelt told the shah. "Prime Minister Churchill has arranged to have a specific change made in the time announcement on the BBC broadcast tomorrow night. Instead of saying 'It is now midnight,' the announcer will say, 'It is now'—pause—'*exactly* midnight.' "

Once all the formal verifications were made, the shah readily agreed to cooperate with Operation Ajax and said he would promptly fire Mossadegh. But the prime minister was tipped off to the plot and went into hiding. The streets of Tehran erupted in chaos, with weeks of fighting between backers of Mossadegh and supporters of the shah. Eventually the shah's army and police won out, Mossadegh was tracked down and jailed, and the country settled into a regime of tyrannical rule until the shah was overthrown by the revolution of 1979.

Western leaders were in agreement that Anglo-Iranian could not just be put back in control of Iran's oil industry. The United States brokered a plan for a consortium of oil companies to take charge. Anglo-Iranian would get the largest share of management, 40 percent; Royal Dutch Shell would have 14 percent; five U.S. companies would each have 8 percent; and the French company Compagnie Française des Pétroles (CFP) would have 6 percent.[6]

Since the name no longer really fit, the Anglo-Iranian Oil Company was changed to British Petroleum in 1954.

As a result of the instability in Iran and other Middle East nations, including Iraq where the British-backed king was beheaded in a 1958 coup, oil companies expanded their exploration efforts in other parts of the world. British Petroleum made its first foray into the Americas through a deal with Sinclair that enabled it to sell gasoline at its U.S. stations and jointly drill for oil in Latin America. The company also acquired some wells in central California—along the Kern River near Bakersfield—through purchases it had made in Trinidad and Canada in 1957. The wells brought in nearly $3 million a year, which BP pumped into its exploration with Sinclair that led to a major discovery in Colombia.

The most promising expansion drive was in Alaska, where geologists were eyeing hilly formations called anticlines on the northern tundra that appeared similar to those in the rich oil fields of the Middle East. British Petroleum geologist Peter Cox, borrowing from his experience in Iran, began the company's exploration efforts on the North Slope in 1959, primarily around the Brooks Range. Five years of drilling ended with no discoveries, so a young Scottish geologist, Jim Spence, took over BP's efforts in 1964 in the Colville River basin. Those wells, too, came up dry, but the company was getting close and didn't realize it. In early 1967, it acquired leases near Prudhoe Bay, focusing on areas around the rim of the anticline at the Put River. A year later, after eight unsuccessful wells, British Petroleum decided to pull up stakes and packed up its drilling rig for shipment elsewhere. But before the rivers had thawed to allow for the rig's transport, ARCO and Humble Oil struck oil at the center of the Prudhoe Bay anticline in March 1968.

British Petroleum quickly reversed course. Convinced that most of the oil would be in the rim areas as it was in Iran's anticlines, it stepped up drilling along the Put River. Within a year, geologists had estimated that the Prudhoe Bay field contained at least ten billion barrels of oil—the largest field in North America—and that about 60 percent of it was within the leases held by BP.[7]

The discovery instantly turned the North Slope of Alaska into America's biggest petroleum-production center, with companies scrambling to drill wells and joining forces to plan an 800-mile pipeline to carry the crude to the port of Valdez on the Gulf of Alaska. The prospects made Alaska the new Wild West, teeming with rugged adventurers hoping to get rich on high-paying but risky jobs in drilling and construction. Into this rough-and-tumble world in November 1969 stepped a slight but energetic apprentice engineer named John Browne, dispatched by British Petroleum to learn the ropes of oil exploration in one of the harshest environments on earth.

Browne was the son of a British war hero, Edmund John Browne, who had gone to work with British Petroleum in 1957. Browne had been an officer with the Black Rats, an elite British tank division formally known as the Fourth Armoured Brigade, which fought in major campaigns across Europe and Africa and helped negotiate the surrender of the Italian Army in 1943. After the war, Browne met a Holocaust survivor, Paula Wesz, in Transylvania, Romania. The two fell in love, were married in 1947, and less than a year later had their only child in Hamburg, Germany.

John Browne, it seems, was greatly shaped by the experiences of his parents. His father's military background showed up in the son's aggressive and risk-taking side, while his mother's horrific tribulations—she had been scheduled for extermination at Auschwitz and lost most of her family in Nazi death camps—made John wary and secretive. He traveled extensively with his parents as a young boy, was sent to boarding school in London at the age of nine, then went to St. John's College, Cambridge, under the sponsorship of British Petroleum, for whom his father worked in both the oil fields of Iran and at London headquarters.

In Alaska, Browne was a fast learner and helpful innovator, moving from work on the drilling rigs to developing computer programs—before computer programs were in vogue—that would help BP solve some of the myriad problems involved with oil production in the Arctic. After two years of virtually nothing but work for Browne, BP managers moved him to New York in 1971 to learn the financial and social aspects of the industry while a lengthy battle was waged in Alaska and Washington over permits for the pipeline project.

BP and three other companies—Standard Oil of Ohio (Sohio), Atlantic Richfield (ARCO), and Humble Oil—had jointly applied for a state permit to build the pipeline in June 1969, but immediately ran into opposition from native Alaskans and environmental groups. There were issues over whether an underground pipe streaming warm petroleum would destroy the tundra's critical permafrost and whether caribou and other roaming wildlife would be harmed by a massive metal structure if it was built above ground. Native Alaskans, meanwhile, were demanding new acreage and monetary payments to settle long-standing rights disputes before they would agree to allow a pipeline to traverse their lands.

It took several years for the litigation and political squabbles to play out. Environmental concerns were addressed by plans to build much of the pipeline above ground, with ramps to allow caribou to wander the tundra uninhibited. Congress solved the land rights problem in 1973 by agreeing to transfer forty million acres and $426 million to native Alaskans. And when the Organization of Petroleum Exporting Countries (OPEC) imposed what became known as the Arab oil embargo on the United States in October 1973, in retaliation for U.S. funding of Israel's military, opposition to the Alaska pipeline dissipated, and the Trans-Alaska Pipeline Authorization Act was signed into law by President Richard Nixon on November 16, 1973.

British Petroleum had been confident the pipeline would be built eventually, and almost as soon as the Prudhoe Bay field was discovered the company began looking for ways to get the oil into the U.S. market. Sinclair, which had passed on the opportunity to explore in Alaska with BP, went on the market in 1969 and was acquired by ARCO, but antitrust concerns forced the merged company to divest some 8,500 Sinclair gas stations, mainly on the East Coast. BP bought the stations for $300 million, but the company's incoming chairman, Eric Drake, believed more U.S. outlets would be needed for the gush of oil from Alaska.

Drake approached Sohio, one of the biggest remainders of the original Standard Oil Company that was broken up by the government in 1911. Sohio had more than 3,500 gas stations, mostly in the Midwest, along with well-established refining and marketing operations, but it was short on petroleum supplies. A merger was worked out with BP initially taking a 25 percent share of the holdings, with an agreement to go above 50 percent once production in Prudhoe Bay reached 600,000 barrels a day by the end of the 1970s. BP would run the Alaska field and work on the pipeline, while Sohio would continue to manage fuel distribution and marketing in the rest of the United States.

The Justice Department moved to block the merger on antitrust grounds, but BP Chairman Drake and Sohio President Charles Spahr scurried to Washington to meet with Attorney General John Mitchell. Their powers of persuasion, combined with pressure from the British foreign minister on Secretary of State William Rogers, led the government to drop its objections provided the new company sold enough service stations to keep its share of the Ohio gasoline market below 25 percent. The merger went into effect on January 1, 1970.

Two years later BP/Sohio bought the chain of Scot service stations, and some of them became the first in America to assume the BP brand. Over the next two decades, BP replaced names like Boron, Gulf, Gas-N-Go, and William Penn across the nation, and even bumped Standard and Sohio out of Ohio, where they had been staples for decades.

British Petroleum also focused on the home front in the 1970s, after capping a fifty-year search for oil in the North Sea with the discovery of the Forties field, a significant deposit under 350 feet of water about a hundred miles off the coast of Scotland. The company invested millions of pounds in an offshore drilling platform and in the largest deepwater pipeline ever laid, enabling it to start pumping 500,000 barrels of oil a day by 1977, or about 20 percent of the U.K.'s needs. The new domestic supply coincided with a severe economic crisis in 1976 that forced the British government to sell 66 million shares of BP stock. Over the next eleven years, the government would divest itself entirely of company holdings, making British Petroleum fully independent for the first time since the outbreak of World War I.

As the company was maturing as an oil major, John Browne was taking more steps up the BP ladder. After a two-year stint in the company's San Francisco office, he returned to the U.K. in 1976 to work on the Forties field. Browne also was sent to Canada to investigate the prospects for recovering petroleum from the oil sands that stretch across the western province of Alberta, but he rejected the idea as too costly and too environmentally destructive.

In 1980, Browne was selected as a Sloan Fellow at the Stanford Graduate School of Business, and he was ordered by BP to accept the prestigious scholarship. After earning his MBA a year later, Browne was named head of BP's exploration and production division in London. There he devised a scheme to sell small shares of

BP oil holdings in the North Sea to companies seeking a British tax break, and ended up adding more than $300 million to company coffers.

Chief Financial Officer Robert Horton rewarded Browne with an appointment as group treasurer and head of BP Finance International, and one of his first tasks was to clean up the financial mess at Sohio, which was bleeding millions of dollars on fruitless exploration efforts. Browne worked up a plan for a Sohio takeover, in which majority shareholders would force changes in the board's makeup and then have the new board recommend a merger. BP Chairman Peter Walters disliked the strong-armed approach, though, preferring to simply replace top officers at Sohio with BP people, like Browne, who would turn around the financial performance. Browne was asked to resign from BP in 1986 and go to Cleveland as Sohio's new CFO.

Browne managed to turn Sohio's negative cash flow to a positive one by early 1987, at which time BP executives in London plotted their own Sohio takeover, leaving Browne in the dark because he was no longer a BP employee.[8] When the deal became final later that year, Browne was named executive vice president and chief financial officer of the new BP America, and was put in charge of the company's North American exploration and production program.

While he was working in Cleveland, Browne became a disciple of a bright young geologist, Jack Golden, who was convinced there were vast oil deposits beneath the Gulf of Mexico. In his new position at BP America, Browne was able to act on Golden's hunch and proceeded to pump the company's entire exploration budget—about $50 million—into acquiring leases and drilling for oil in the Gulf.

A BP-Shell partnership tapped into the Mars oil field about 130 miles southeast of New Orleans in 1989, and a series of other major discoveries soon followed. By the end of the 1980s, BP America was the largest oil producer in the United States.

BP now began to take on more of the characteristics of Big Oil, at least as it is perceived by much of the public. When the *Exxon Valdez* smashed onto the rocks in the Gulf of Alaska on the night of March 24, 1989, unleashing 11 million gallons of crude oil, it was BP that was supposed to lead containment efforts as the controlling owner of Alyeska, a consortium of Alaskan oil interests. But it took more than twelve hours for Alyeska, taking orders from BP headquarters in Houston, to get equipment to the site of the spill, and the barge arrived without enough booms and with no containers to hold oil once it was skimmed from the surface. As a result, a spill that could have been confined to a relatively small area spread more than a thousand miles along the coastline, according to an Alaskan commission that investigated the *Valdez* disaster.

In the same year, BP America pushed hard for the United States to allow drilling in the Arctic National Wildlife Refuge (ANWR), which the industry believed could contain millions of barrels of oil. While environmental groups were extolling the virtues of preserving one of America's last true wilderness areas, Roger Harrera, a top official in BP's Alaska division, dismissed their claims as "an issue of local fauna." The environmentalists won the battle in Congress, keeping legislation that would have opened ANWR to drilling bottled up in the Senate.[9]

On the world stage, too, BP developed a reputation for putting profits over social justice. After a coalition of nations, including the

United States, imposed trade sanctions on South Africa to protest its apartheid system of racial separation, BP circumvented the oil embargo by shipping crude into the country through a series of "chartering schemes," according to a United Nations report in 1988. "While small companies and middlemen were reported as being the embargo's main violators, the role of major transnational oil companies which owned subsidiaries in South Africa—such as British Petroleum (BP), Caltex, Mobil, Shell, and Total—could not be underestimated," the *UN Chronicle* reported. A decade later, after Nelson Mandela had risen from bondage to become South Africa's president, a Truth and Reconciliation Commission he had appointed to investigate human-rights violations under apartheid roundly condemned the mining and oil industries for supporting the old regime. BP was included in a list of companies cited by the commission as among "the most notable" for refusing to cooperate with its investigation.

John Browne had a bird's-eye view of BP's maneuvers through global politics, as he rose to CEO of BP Exploration in 1989, to the BP board as a managing director in 1991, and to BP's Group CEO in 1995. Browne wrote in a 2010 memoir that his experiences in Alaska, in Cleveland, and in other parts of the company hierarchy made him worry about the oil industry's image.

Browne said one of his deputies, Rodney Chase, told him that employees would not have a good incentive to perform if the goal was simply boosting shareholder value. "He was right," Browne wrote. "If we were seen as dirty, old-fashioned, and short on ethics or environmental principles, we would not be able to attract the brightest and best of the next generation."

Shortly after becoming BP's CEO, Browne laid out his concerns about the industry's public profile at an oil and gas conference in Houston. "They say the industry damages environments and communities and that it generates pollution and waste," he told energy business leaders. "They say those involved in the industry are secretive, arrogant, and driven solely by profit."

The following year, Browne made a bold move toward trying to change that image. In a May 19, 1997, speech at Stanford University, he stunned his peers in the oil industry by declaring that BP was taking seriously the growing scientific warnings about climate change. "I believe that we've now come to an important moment in our consideration of the environment: the moment when we need to go beyond analysis to seek solutions and to take action. It is a moment for change and for a rethinking of corporate responsibility," he said.

"The time to consider the policy dimensions of climate change is not when the link between greenhouse gases and climate change is conclusively proven, but when the possibility cannot be discounted and is taken seriously by the society of which we are part," he said. Browne went on to outline steps BP would take to address the problem, including a program to cut the company's carbon dioxide emissions, development of a voluntary emission-trading system proposed by the Environmental Defense Fund, investments in solar technology and other forms of clean energy, funding for climate research and policy debates, and participation in a Nature Conservancy project in Bolivia aimed at buying up logging rights to preserve forests that can capture carbon.

Browne's speech was the product of years of consultation and debate both inside and outside the company, according to 1997

interviews with BP executives by industrial ecologist Ernest A. Lowe
and SAIC consultant Robert J. Harris.[10]

Publicly, BP had kept stride with the rest of the oil industry in
the early 1990s and maintained its membership in the Global Climate Coalition, a lobbying group formed in 1989 to oppose an international treaty on climate change. But inside the corporation's
executive offices, attitudes had begun to change. Robert Horton,
who had risen from BP's CFO to CEO in 1990, had the company
participate in the Earth Summit in Rio de Janiero in 1992, with BP
executives sharing the stage with soon-to-be vice president Al Gore,
a leading advocate for action on climate change. BP also organized
a series of forums in the United States and Europe, inviting leaders
of environmental groups, including Greenpeace, to provide unvar-
nished assessments of the company's performance in closed, off-the-
record meetings. The sessions were chaired by the late John Sawhill,
then the president of the Nature Conservancy, although the powerful conservation group was not an official sponsor.

Sawhill, who died in 2000, credited Browne with putting the
input from those meetings into a company action plan. "I've known
Browne for twenty years and he does not like to run with the pack,"
Sawhill told Harris in September 1997. "Browne sees BP's position
and his speech as a way to distinguish them from others in their
industry."

There was also a clear financial incentive behind the strategy,
Sawhill said. "BP wants to redefine the way business is done in their
industry. Make the new rules," he said.

BP executives echoed that theme to Lowe and Harris. "BP has a
desire to be constructively engaged in all of the issues of the climate

issue, in the areas that impact our global business," one said, according to notes kept by the two researchers. Another pointed out that the primary job of the oil industry is "providing mobility for people," and if a new, cleaner source of energy, such as fuel cells, proved to be the wave of the future, BP needed to be riding that wave.

"We suffered a downturn like many companies in '92, and it became a crisis for us," another executive said. "Our '92 financials were dramatically bad and that triggered a sea change in how BP viewed its operations. We took a lot of steps to refocus and became a much flatter organization." One step was the development of "strategic alliances" with environmental groups and others who would be viewed as the enemy by most oil companies. And there were some clear economic reasons for doing so.

One of the leading environmental groups in Europe, Greenpeace, had begun arguing in the early 1990s that more violent storms and floods caused by global warming would take a severe toll on insurance companies as damage claims rose. Five major European insurers took notice, and began letting corporate customers know that environmental performance would be considered when setting premiums. A number of European nations also began moving toward limits on greenhouse gases and stepped up pressure on the U.S. government to follow suit.

A tipping point for BP's groundbreaking stance on global warming may have come in February 1997, when more than 2,500 members of the American Economics Association, including eight Nobel laureates, issued a statement urging public policy to be "guided by sound economics rather than misleading claims put forward by special interest groups." The statement concluded: "As economists, we believe that global climate change carries with it significant environmental, economic, social, and geopolitical risks, and that preventive steps are justified." A month later Browne made his speech at Stanford, and BP made its break with the industry on the issue by withdrawing its membership in the Global Climate Coalition. His peers in the industry scoffed at Browne's move. Five months after the Stanford speech, Exxon CEO Lee Raymond derided calls for action on climate change in a speech to the World Petroleum Congress.

BP's steps toward greener pastures capped what had been a roller-coaster financial stretch for the company. Chairman and CEO Horton started the decade with Project 1990, a sweeping plan for cutting costs by simplifying and decentralizing the corporate structure. Mostly the effort slashed jobs—more than 19,000 of them, or 16 percent of BP's workforce—and lowered the morale of overburdened employees who survived the layoffs. Horton was forced to resign in 1992 when, despite all the savings, BP projected an $811 million loss for the year after sales revenues dropped from $66.4 billion in 1990 to $51.9 billion in 1992, largely due to the global recession. Horton was succeeded as CEO by David Simon, a veteran marketing and financial manager who also cut costs but focused more on encouraging better performance among the BP staff. By 1995, when Browne took over as CEO and Simon became chairman, BP was paying record dividends off rising profits. Still, in Browne's own words, when he became CEO on June 10, 1995, BP was spinning its wheels as a "middleweight insular British company."[11]

Browne believed BP needed to expand its natural gas business and move more aggressively into growing markets such as Asia if it was going to become a truly global player. With Simon's support, he courted Mobil as a merger partner in late 1996 and early 1997, but the U.S. company's CEO and chairman, Lou Noto, resisted. So Browne turned his attention to Amoco, which had extensive gas reserves and a large U.S. refining and marketing operation. Negotiations took more than six months, but in August 1998 the marriage was announced: BP would acquire Amoco for $48 billion. It would be the largest industrial merger in history to that point, with the two companies worth a combined $110 billion.

The threat of the BP-Amoco powerhouse forced other oil companies to follow suit in a wave of industry mergers: Exxon joined with Mobil, Chevron with Texaco, and Conoco with Phillips. BP-Amoco upped the ante in 1999 by acquiring ARCO for $25 billion, though antitrust issues raised by the government forced ARCO to sell some of its Alaska assets first to prevent too much concentration of the state's energy industry in one company. Browne moved immediately after both the Amoco and ARCO acquisitions to eliminate duplication and inefficiencies in the new company, setting a goal of eliminating nearly 15,000 jobs within a year or two and reducing costs by $4 billion within three years. By the end of 1999, the combined corporation had been slimmed down from more than 100,000 employees to fewer than 90,000.

The kudos Browne had received from his dramatic announcement on climate change led him to expand that effort, too. He asked one of New York's top marketing firms, Ogilvy Public Relations Worldwide, to help "rebrand" the company with a cleaner, more progressive image. The resulting Beyond Petroleum campaign would become one of the most successful green marketing programs in history.

The campaign served a number of useful purposes. The new logo that was developed, named for the Greek sun god Helios, featured a yellow sun on a green background with the lowercase letters *bp*, giving the company a softer, brighter image. The slogan "Beyond Petroleum" made BP seem focused on the future, and deliberately took "British" out of its name to avoid reminding Americans it was a foreign company. And the suggestion that it was moving past dependence on oil allowed BP to give the impression it was no longer part of an industry considered by most to be decidedly unfriendly to the environment.

The Beyond Petroleum campaign was not without its critics and skeptics. Many environmentalists considered it to be nothing more than expensive "greenwashing." Others viewed it as merely annoying. *New York Times Magazine* writer Darcy Frey described it this way in a December 2002 story (making full use of the lowercase style that bp adopted): "New Yorkers in particular were the target of a high-saturation ad campaign that felt, at times, like an overfriendly stranger putting his arm around you in a bar. In Times Square, a huge billboard went up, reading, 'if only we could harness the energy of new york city.' Then the stranger, perhaps feeling the need to explain his intentions, went on: 'solar, natural gas, wind, hydrogen. and oh yes, oil.' Finally, the stranger took his arm away with a bit of a shrug: 'it's a start.' "[12]

The campaign was launched a few days before the September 11, 2001, terrorist attacks, then was withdrawn for many months while America mourned the tragedy of that day. But by the time Beyond

Petroleum had been revived, run its course, and was put to bed later in the decade, it had clearly instilled a new image in the minds of many. One informal survey conducted in 2008 found that BP was identified as a green organization more frequently than one of the pioneers of the environmental movement, Greenpeace.

John Browne and BP were on top of the world entering the new millennium. The company's sudden stature as one of the world's biggest energy giants prompted Queen Elizabeth to make Browne a British knight in 1998, and a few years later he was named to the House of Lords as a crossbench, or independent, member. Lord Browne of Madingly became a major supporter of the Royal Opera House and London's National Gallery; he was invited to parties hosted by pop star Elton John and actress Gwyneth Paltrow; he regularly held court with political leaders like Dick Cheney and Tony Blair. He was even celebrated by *Vanity Fair* magazine as one of the world's top green leaders, a remarkable accolade for the head of a mammoth oil company.

chapter three

THE THREE PIGS

AFTER THREE WORKERS DIED IN 2004 at BP's Texas City refinery southeast of Houston, site manager Don Parus started digging into records for the sprawling, 1,200-acre complex that processes some 10 million gallons of gasoline every day—nearly 3 percent of the U.S. supply.

Parus, just recently hired at the plant, was stunned to find that in the previous thirty years, twenty other employees were killed in accidents, a rate of one every eighteen months.

Two of the deaths occurred on Parus's watch, on September 2, 2004, when superheated water blew out of an overpressurized pipe and fatally burned two workers. Parus later broke down in the court-room when he recalled visiting one of the workers at the hospital and learning a short time later that the man had died. The accident resulted in a $109,500 fine for eight violations cited by the federal Occupational Safety and Health Administration (OSHA), but little was done to upgrade aging equipment at the plant that was taking a heavy toll on the workforce.

Parus was concerned enough to hire a workplace safety consultant, the Telos Group, to conduct a survey of the refinery's 1,800 employees. The damning results that came back a few months later were as disturbing as the death statistics. Workers who were guaranteed anonymity gave a harsh assessment of plant operations in comments describing faulty and antiquated equipment, inadequate staffing, sloppy communications, inattention to safety requirements, pressure to produce profits, and managers focused on cost-cutting above all else. Among the hundreds of responses to the survey:

> The short staffing in maintenance and operations is a big issue; we have cut too thin.

> The equipment is in dangerous condition and this is not taken seriously. At the refinery there's a frame of mind like "we are the ones that make the money"—they take pride in running on thin air, but if they do it by killing someone every eighteen months then you don't have bragging rights about production.

> There are lots of issues about the effectiveness of alarms.

> Pipe thinning worries me the most; its failure could be catastrophic with little warning.

> We have big issues with piping. . . . We're just now getting our inspection teams out and the data coming back is very scary.

> It seems like it all comes down to money. We tell them we need it. They tell us they don't have the money. As soon as it blows up or someone gets hurt there's all sorts of money.

> Units are 90 percent of the time run to failure, due to postponing turnarounds. So making money or saving money for that particular year looks good on the books.

What pleases managers most is when you reduce costs. It doesn't matter if you cut corners that shouldn't be cut as long as you reduce cost.

When executives from London come, it's all about us trying to survive their intimidation with our presentations, etc. It never seems to occur to anybody to ask, "What do we need from them? How are they doing with their accountabilities to us?"

The history of investment neglect coupled with the BP culture of lack of leadership accountability from frequent management changes is setting BP Texas City up for a series of catastrophic events . . . this place is nearing an investment requirement on the scale of $450 to $500 million.

The consultants from Telos summed up their sixty-two-page report with a bleak observation: "We have never seen a site where the notion of 'I could die today' was so real for so many hourly people."

It was no wonder the workers at the nation's third-largest refinery feared for their lives. The plant, in operation since 1934, had a long history of problems and stood in the blast zone of what is widely considered the worst industrial accident of all time in America. On April 16, 1947, a French cargo ship in the Texas City harbor was being loaded with fertilizer to assist farmers in postwar Europe when a fire broke out that caused an explosion felt in Houston, more than thirty miles away. The blast set off a chain reaction of other fires and explosions, including one on another ship in the harbor carrying fertilizer, and by the time the dust settled more than 560 people were dead or missing. Coincidentally, in that same year, Amoco acquired the Texas City facility.

At the refinery itself, there were serious safety issues dating back to at least 1992, when a leak of flammable liquids formed a vapor cloud that exploded and burned a worker named Guy Holdren, who died fifteen days later.[1] OSHA later cited Amoco, the plant's owner at the time, for operating with a faulty piping system and recommended installation of a new venting system that would burn off hazardous gases. The flaring equipment was never installed.

There were numerous other explosions at the plant, including one in 1995 that put more than a hundred people in the hospital. Three Texas City workers died after BP took over the plant with its purchase of Amoco in 1998.

Parus took his data on deaths to top executives at BP, and no one seemed to care. "It didn't shock anybody like I thought it would," he would testify in court several years later.

Instead, Parus was handed a challenge at the beginning of 2005: Find a way to cut more than 20 percent—at least $65 million—from the Texas City refinery's operating budget of $300 million. The plant had already been ordered to make across-the-board spending reductions of 25 percent when BP bought Amoco, and there were cuts of at least that amount in the 1990s as Amoco struggled to become a leaner corporation. Parus said he "pushed back" as much as he could, but the best he could get from headquarters was an agreement to reduce the cost-cutting goal to $48 million.

Officials at BP headquarters knew they could count on Parus to meet the target. "The prevailing culture at the Texas City refinery was to accept cost reductions without challenge and not to raise concerns when operational integrity was compromised," the company's vice president of group refining wrote in a 2002 report. Parus clearly knew what was expected of him, even though he was fairly new to Texas City and was its sixth plant manager in five years. Early in 2004, he and other south Texas managers were shown a

PowerPoint presentation outlining how cuts in equipment and monitoring costs had saved the company more than $100 million at Houston-area plants. The managers were encouraged to stay on that track. And at the end of the year, a BP regulatory affairs manager, Susan Moore, was nominated for a $1,000 bonus for successfully lobbying to exempt Galveston County, where Texas City is located, from a new state cap on emissions of volatile organic compounds. The exemption reduced BP's pollution-control costs by at least $150 million, Moore's supervisor wrote.

There was also a long-held principle at BP that the value of a human life should be a part of every cost-benefit analysis that was done for calculating appropriate budgets for company operations. Shortly after BP bought Amoco in 1999, BP engineer Robert Mancini lamented in a memo that the new acquisition's management did not consider human-life costs in making its risk assessments and budget analyses. "Amoco was generally unwilling to take this step," Mancini wrote. "This is more a cultural issue than a technical one, but one that will have to be addressed."

In a later memo, Mancini made an apparent attempt at lightening up a macabre subject by drawing up a chart showing a cost-benefit analysis of how different types of structures would protect workers from an explosion. At the top of the chart, just below the BP sun logo, were drawings of three pigs soaked in oil, with slots on their backs to make them look like piggy banks. Droplets of oil were popping up from the slots. The analysis then assessed the vulnerability of houses made of straw, sticks, bricks, and "blast-resistant" materials, and estimated the value that would be lost if "the big bad wolf" attempted to blow each one down.

Below the chart, in a section entitled "Cost benefit analysis of three little pigs," Mancini explained that the value of a piggy life would be $1,000. Then he asked, "Which type of house should the piggy build?" In the chart, the straw house had a vulnerability rating of 0.9, giving the pig a one-in-ten chance of survival if the wolf tried to blow it down and putting the value of the expected loss at $900, or nine-tenths of $1,000. Since the straw house would cost only $10, the total expected loss would be $910, the chart showed. The evaluation for the house made of sticks showed a vulnerability rating of 0.8, an expected value loss of $800 and a construction cost of $20, for an expected total loss of $820. For a house of bricks, the vulnerability rating was 0.1, the value loss was $100 and the building cost was $100, making the expected total loss $200. And for the "blast-resistant" house, the vulnerability was 0.01, the value loss was $10, and the building cost was $1,000, for a total expected loss of $1,010. A handwritten note on the chart said "optimal" with an arrow pointing at the brick house, since the total estimated loss would be only $200, though the risk to the pig would be ten times greater than in the blast-resistant structure.

The cold algebra of costs and benefits had long been tipping against the workers in Texas City, which one BP official described as "the most complex facility in the world." Parus, the refinery manager, appeared caught up in the culture when he sent an e-mail to his BP bosses on March 17, 2005, saying, "The Texas City site undoubtedly experienced the best profitability ever in its history last year," with profits of nearly $1 billion recorded at the refinery in 2004. Parus boasted that safety performance so far in 2005 was "maybe the best ever," but there was still concern among the workforce "over the condition of our assets and infrastructure."

Other managers at the plant were much darker in their descriptions of conditions at that time. "I truly believe we are on the verge of something bigger happening," a plant safety manager wrote in a memo on February 20, 2005. Three weeks later a BP business plan for the refinery listed key safety risks, including, "TCS [Texas City Site] kills someone in the next twelve to eighteen months."

It took only eight days for the fear to be realized.

In the early morning hours of March 23, 2005, the night shift staff was preparing for the restart of the refinery's isomerization unit, which distills highly flammable hydrocarbons such as pentane and hexane to make the additives used to boost the octane of gasoline. The unit had been down for a maintenance overhaul, or turnaround, and the process of restarting such high-temperature, high-pressure equipment was a delicate task, one of the most dangerous operations at any refinery.

Workers began by pumping volatile liquids into a "raffinate splitter" tower where the distillation process is done with a furnace that vaporizes the hydrocarbons. The tower is 164 feet tall, but it is critical for the liquids at the bottom to be kept at a level between five and ten feet—if the liquid drops below five feet, the furnace can be damaged; if it goes above ten feet, pressure inside the tower can rise to extreme heights. There are three important devices that enable workers to determine how much liquid is in the tower. One is a level indicator that shows 0 percent if the liquid is below three feet and 100 percent if it is above ten feet; the optimum reading should be around 50 percent. Then there are two alarm systems: one that sounds when the liquid drops below three feet or rises above ten feet, and one that sounds if the liquid rises high in the tower and reaches an overflow drain. Tragically, none of these devices was working properly on that Wednesday morning.

At 5:00 A.M., after starting the process of pumping liquid into the tower, the lead operator in the unit left the plant an hour before his shift was scheduled to end. His replacement arrived at 6:00 A.M. and began a twelve-hour shift for the thirtieth day in a row. The day shift supervisor arrived at 7:00 A.M., one hour past his scheduled start time. Neither the lead operator nor the day supervisor had received much information from the night shift staff, other than a scribbled note saying that some raffinates had been injected for the restart of the "isom" process.

Shortly before 11:00 A.M., the day supervisor left to attend to a family emergency, leaving a single operator on duty to monitor three different control boards, including the one for the isom unit. Budget cuts in 1999 had eliminated the position of the second board operator in the unit. The readings on the isom unit's control board gave no indication of any problems, with the level indicator showing the depth of the liquid in the tower to be hovering between eight and nine feet. In reality, the liquid had risen to nearly a hundred feet, with no alarms sounding to indicate the hazardous situation.

Just outside the isom building, several hundred contract workers were gathering to head for a company luncheon celebrating a month without any lost-time injuries. The workers had been set up with temporary offices in trailers parked between a hundred and a hundred and fifty feet from the isom unit, including a double-wide directly across the street from the plant that was used by employees of J. E. Merit Constructors, a Texas firm that was part of the engineering company Jacobs Field Services. None of the contract workers, even while gathering for a safety event, had been told that the risky process of restarting the isom unit was under way.

Inside the plant, readings on the control board at 12:41 P.M. showed a buildup of pressure inside the tower. Uncertain of why that was happening, workers lowered the temperature of the furnace and opened a "blowdown drum" to vent gases from the tower, while the board operator opened a valve that would allow excess liquid to flow out into a holding tank. But inside the tower, the chemical processes were boiling over and getting out of control. The liquid had risen above 158 feet, even though the level indicator showed it was around eight feet, and hot gases mixed with bubbling liquids began to overflow. As the holding tank at the base of the tower began overflowing into the sewer, flammable gases and liquids began shooting out a vent at the top of the tower as well. The time was around 1:20 P.M. Within ninety seconds a vapor cloud formed over the entire plant and spread over the contractors' trailers nearby, just after most of the workers had returned from their lunch and were settling back into their jobs.

Two workers were sitting in their pickup truck just outside the isom plant when the engine started to race as the flammable vapors seeped inside the running engine. Unable to turn off the engine with the ignition switch, the workers jumped out of the truck and began to flee. As they did, they heard the engine rev up and backfire, apparently causing a spark that ignited the gas. An unbelievably powerful explosion occurred—one nearby worker later described it as ten times the sound of loud thunder—and a fireball enveloped the isom plant. The flames roared downward and flattened the J. E. Merit double-wide and an adjacent trailer, then triggered as many as five more explosions, including several vehicles parked near the trailers.

Ralph Dean had just left his wife, Alisa, at her office in the Merit trailer and had resumed his work on a forklift about two hundred yards away when the initial blast occurred. He turned and watched a second explosion obliterate the double-wide. Dean was blown off of his forklift, but jumped up and began running toward the remains of the trailer. "Everything was on fire, and the trailer was flat," Dean, then forty-five, would tell *The Houston Chronicle* eight months later. "There were all kinds of things coming out of the air. Balls of fire. Pieces of pipe. Wood. Pure destruction. It was like someone had made a bomb run on us."[2]

When Dean reached the trailer, he heard the screams of survivors buried beneath the rubble and he started removing debris to try to reach them. Just then, a nearby truck exploded and flames leaped toward the demolished trailer. Dean, fearing survivors would be burned alive, ran to get his forklift and pushed the burning truck away. As other vehicles started to blow, he cleared the entire parking lot with his forklift before rushing back to the trailer site.

By this time other rescuers had arrived, and Dean and another man found BP worker Jack Skufca lying on top of another survivor. Skufca's chest had been crushed by a piece of office equipment, but he was still alive. After helping pull Skufca out of the rubble, Dean found a dead man he recognized immediately. It was his wife's father, Larry Thomas, who also worked for Merit.

Dean now became almost maniacal in a quest to find Alisa. He stopped to help free two more workers, Kristof Harris and Andy Mc-Williams, but thought momentarily about leaving them. "I wanted to let them die," Dean confided to the *Chronicle*. "God save my soul, I just wanted to find my wife. But I couldn't do it."

Finally Dean did find Alisa beneath a broken bookshelf, burned and suffering from injuries that would keep her hospitalized for four months, including a cracked skull and a broken neck and ribs. He helped get her to a rescue helicopter and returned to search for more survivors.

By now the scene was tantamount to a war zone, with emergency vehicles and helicopters shuttling in and out to fight the fires, attend to survivors, and transport more than 180 injured victims to hospitals.

Harris, one of the men Dean helped rescue, later said there was no logical reason he did not die in the explosion. Instead he was miraculously buried under several feet of rubble that gave him air to breathe but shielded him from the fire. "I can't feel lucky that I lost so many friends," he told the *Chronicle* later. "Too many good people lost their lives."

Among them was Ryan Rodriguez, twenty-eight, who was found by his friend and mentor, Pat Nickerson, lying in the rubble with his face torn to bits by flying debris. Nickerson, fifty-five, watched the young man die and was tormented about it for months afterward. "I thought about why I wasn't severely injured or even killed," Nickerson wrote in a diary he shared with the *Chronicle*. "Ryan was just twenty-eight years old and . . . he had everything ahead of him. Why him and not me?"

Others who were killed in the explosion or died shortly afterward, in addition to Larry Thomas and Ryan Rodriguez, were Kimberly Smith, Larry Linsenbardt, Morris King, Daniel Hogan, Eugene White, Rafael Herrera, Glenn Bolton, Jimmy Hunnings, Susan Taylor, Lorena Cruz-Alexander, and Arthur Ramos.

And there were James and Linda Rowe, a husband and wife from Hornbeck, Louisiana, who opted to take short-term jobs with the same contractor rather than live 164 miles apart for a months-long stretch. The couple had already done that for much of their lives. So after working happily together since October, the Rowes died horribly together in the Merit trailer: Linda was decapitated by flying debris, and James was so bloodied on his face and skull that it is assumed he died almost instantly in the mammoth, fiery blast.

Eva Rowe was on her way from Hornbeck to Texas City on the afternoon of March 23, looking forward to an Easter weekend with her parents, whom she regarded not just as mother and father, but as her closest friends in the world. Eva was twenty at the time, still searching for a path to take from what had been a joyful family life but a troubled adolescent period, what with little else to do but party in the rural woodlands of southwest Louisiana. She had begged her mother to stay with her in Hornbeck while James was working in Texas, even trying half jokingly to scare Linda about refinery work, which Eva had experienced briefly at a Corpus Christi plant where her father had found her an office job while he was a contract worker there. "Mom, if you go there, you will die there," Eva said she told Linda shortly before she left.[3]

About forty-five minutes from Texas City on State Highway 146, Eva stopped for gas and went inside the station to pay. For no apparent reason the attendant looked at her and said, "Everything is going to be okay, my friend," at which point Eva noticed the TV showing huge billows of smoke rising from the BP refinery in Texas City. She pulled out her cell phone and dialed her mother's number, which Linda almost always answered. Nothing. She dialed again and again. No answer. A sinking feeling of despair engulfed Eva Rowe.

As she continued to drive toward the city, Eva could see the smoke even though it was still miles away. She went to a relative's house near Galveston Bay, and could see the refinery across the water with flames still shooting skyward. She watched the news for a while on TV, then went to her parents' trailer hoping against hope that they had escaped and gone home. No one was there. It wasn't until late that night at her relative's home that her worst fears were realized. She went to the home of a neighbor who worked at the plant. When he opened the door and saw Eva, the man dropped to his knees and broke down. Rescue workers had not found James or Linda.

Around 4:30 A.M. on March 24, Eva finally reached the Merit office and was told that both of her parents were dead. She should go to the convention center in Texas City and help identify the bodies. Eva went there, but couldn't look at her parents' remains. She was given a photo that she stuffed in her pocket without a glimpse. She knew it was James and Linda; and she knew that at that moment her own life as she knew it had ended, and that she had to start all over from scratch.

BP had immediately taken responsibility for the tragedy and a stunned CEO John Browne rushed to Texas City to show his support the day after the accident. "We have a very simple rule at BP—that we are responsible for what happens inside the boundaries of our plant. This is no exception. We will be doing everything we can to assist the families," Browne said in a Thursday afternoon appearance at city hall. He added that in his thirty-eight years with BP, March 23, 2005, was the darkest day of his life.

Browne tried to assure Americans that the shutdown of a small portion of the refinery would not disrupt the nation's fuel supply. "The bulk of the refinery is operating, and operating well," he said. But that afternoon the price of gasoline shot up nearly 3 cents on world markets, pushing average U.S. prices to just over $2 per gallon. It was the beginning of an upward trend that would continue until mid-2008, when prices peaked at $4 a gallon before the start of a recession that would bring them back to pre-2005 levels.

BP America President Ross Pillari pledged that the company would thoroughly investigate what happened at Texas City. "It's clear that we have a lot of work to do in the coming days to make sure we understand exactly what happened, and we're going to do that," Pillari said the day after the accident. "We are going to put all of our resources into it."

It wouldn't be revealed until much later, but a pair of e-mails sent by top BP officials on those dark days undercut the sincerity of BP's public statements. John Manzoni, BP's chief refining officer who had accompanied Browne to Texas City, sent a message to a friend on March 24 that subsequently surfaced in court documents. "I arrived in Texas City at 3 A.M. along with Lord Browne," Manzoni wrote. "And we spent a day there at the cost of a precious day of my leave."

Late on the day of the accident, the head of BP's public relations team, Patricia Wright, sent a message to all the company's mouthpieces expressing hope that the crisis would soon pass as the Easter weekend approached and the press moved on to other stories. "Media coverage has been very heavy—looks like injuries and loss of life are heavy," Wright wrote to her colleagues. "Expect a lot of follow-up coverage tomorrow. Then I believe it will essentially go away—due to the holiday weekend."

Wright closed with a reference to the ongoing coverage of the dying Terri Schiavo, a Florida woman who had been in a coma for fifteen years and whose life support was cut off on March 18, 2005, after a lengthy court battle between her husband and her parents. "This is a very big story in the U.S. right now," Wright said of the refinery blast, "but the Terry Schiavo story is huge as well." When Wright's message was disclosed during litigation later, one of the attorneys for Texas City accident victims said it made him "sick to my stomach."

Investigators for the U.S. Chemical Safety Board descended on the accident scene almost immediately, as did officials from other federal and state agencies with oversight responsibilities for the refining industry. The board's Angela Blair said information about what was going on inside the refinery at the time of the blast was pouring in from workers. At the same time, attorneys for survivors and the families of victims began filing lawsuits against BP demanding billions of dollars in compensation. Less than two weeks after the explosion, Galveston County State District Judge Susan Criss handed future plaintiffs an early victory by ordering BP to quickly provide documentation about the accident, including all computer records from the isomerization unit where the event started. The company soon set up a $700-million fund to pay claims and promised it would move swiftly to settle cases and avoid lengthy trials.

Gradually the story line began to unfold that BP had been neglecting maintenance and skirting health and safety mandates, and not just in Texas City. *The Houston Chronicle* reported in May that the company had more deaths at its five U.S. refineries in the previous ten years—twenty-two since 1995—than any other refining company. The next highest death toll was half that total, the eleven workers killed at refineries owned by Shell Oil Products, its predecessor Equilon Enterprises, and its sister company Motiva Enterprises. ExxonMobil, the largest U.S. oil company, had only two deaths in the previous decade and none at its two biggest refineries.[4]

Environmental issues at BP refineries also came into sharper focus. Just five days before the Texas City explosion, BP had agreed to pay $81 million, including $25 million in penalties, for a decade's worth of toxic air pollution violations at its Carson, California, refinery south of Los Angeles. BP admitted no wrongdoing, but at the time the settlement was the largest ever for a single facility under the federal Clean Air Act, and suggested that BP had been illegally exposing millions of people to hazardous pollution for years. It was hardly an approach that fit the company's friendly, green image.

Pillari, the president of BP America, followed up his appearance with Browne the day after the accident by holding a press conference in Texas City on May 17 to announce interim results of the company's investigation of the accident. The findings were something of a stunner, although maybe not to refinery employees. BP laid the blame squarely on the shoulders of the six workers responsible for restarting the isom unit. "If isom unit management had properly supervised the start-up or if isom unit operators had followed procedures or taken corrective action earlier, the explosion would not have occurred," Pillari said.

The employees were even blamed indirectly for the explosion's devastating impact on trailers parked less than 150 feet from the isom unit, when industry standards require temporary offices to be located at least 350 feet away from operating refinery units. "The decision to place the trailer near the blowdown stack was preceded by a hazard review that did not recognize the possibility of multiple failures by isom unit personnel could result in such a massive flow of liquids and vapors to the blowdown stack," Pillari said.

"The mistakes made during the start-up of this unit were surprising and deeply disturbing," he said. "The result was an extraordinary tragedy." Even when it was clear that something was going wrong in the unit, the operators "failed to activate evacuation alarms, denying other workers the opportunity to get out of harm's way," Pillari concluded. "We cannot ignore these failures. For that reason, we have begun disciplinary action against both supervisory and hourly employees directly responsible for operation of the isomerization unit on March 22 and 23."

The six employees—later identified as operators Steven Adams, Andy Tenhaaf, and Warren Briggs, and supervisors Scott Yerrell, Larry Davidson, and Charlie Logan, when they filed a libel suit against BP—were fired, but the company instantly found itself in the middle of another explosion. The United Steelworkers union, which represents Texas City refinery workers, charged that the employees were being made scapegoats for management's negligence on safety, and vowed to pursue its own investigation of the accident. Attorneys for victims of the blast accused BP of using a "public relations stunt" to shift blame away from corporate headquarters.

Just eight days after Pillari's press conference, the company backed off. "We simply used the wrong language to describe the report's findings," BP spokesman Hugh Depland said. "Our fault." The real cause of the accident still had not been determined, he added.[5] All six workers who were fired later settled their cases against BP.

A month later, the Chemical Safety Board's lead investigator, Don Holmstrom, had his own press conference to give an update on his findings so far. "Specifically, investigators from the U.S. Chemical Safety Board have found evidence that several key pieces of process instrumentation malfunctioned on the day of the accident," he said. "Alarms that should have warned operators of abnormal conditions in the isomerization unit did not go off."

Holmstrom emphasized that the "root cause" of the accident had not been determined—that was a decision to be made later by the five-member safety board once all the evidence was gathered and a report was prepared. But it was becoming clear that employees should not bear the brunt of the blame.

Safety board investigators had to scurry back to Texas City in July when another explosion at the refinery rattled the community. Gloria Randle, who was sitting down to watch *Wheel of Fortune* in her home less than a mile from the refinery, was shaken from her chair by a thunderous blast. "I thought al-Qaida was here. I did, I'm not going to lie," she said.[6] The blast and subsequent fire caused no injuries, and was blamed on the wrong type of pipe being installed in a line containing flammable gas, but it put the spotlight on safety issues in the plant just four months past the March tragedy.

The July 28 explosion occurred in a unit that removes sulfur from crude oil, and it turned out the operation had twenty-two safety incidents in 2004—more than any other unit in the refinery—including one that was a major event that "could have been more severe," according to the Texas Commission on Environmental Quality. The reports on the explosion also unearthed evidence that the U.S. Environmental Protection Agency was investigating possible air pollution violations at the BP refinery. "If you are comparing them to other facilities in Texas City, they do have a high number of incidents," said Ronnie Schultz, director of environmental health programs for the Galveston County Health District, which tracks local pollution complaints.[7]

BP announced in August that it planned to install flaring equipment to burn off vapors emitted from vent stacks in all of its refineries, such as the one that overflowed in Texas City on March 23. The replacement plans had been in the works for more than a decade, but had been delayed at least twice by budget considerations. The move did nothing to ease growing concerns about the company at the Chemical Safety Board (CSB), though. On August 18, CSB Chairwoman Carolyn Merritt took the unusual step of issuing an "urgent safety recommendation" in the midst of its on-going investigation, calling on BP to set up an independent com-mission to take a hard look at the company's safety programs.

CEO Browne issued a statement from London that said BP would comply with the request. "The Texas City explosion was the worst tragedy in the recent history of BP, and we will do everything possible to ensure nothing like it happens again," Browne said. "To-day's recommendation from the CSB is a welcome development, and we take it seriously."

BP turned to a veteran Texas lawyer and statesman, James A. Baker III, to take on the assignment of thoroughly reviewing the company's "safety management and culture." Baker, who was secre-tary of state under President George H. W. Bush from 1989 to 1992, would lead a ten-member panel made up of engineers, consultants, researchers, and a former Republican senator—Slade Gorton of Washington—to conduct the study and report back in about a year.

Regulatory hammers started coming down on BP in September, when the U.S. Occupational Safety and Health Administration (OSHA) announced the company had agreed to pay more than $21 million in fines for the March accident. "We know this settlement can never replace the lives that were lost or comfort the families that were devastated by this tragedy," said Jonathan L. Snare, act-ing assistant secretary of labor for OSHA. "But the agreement means that BP Products employees will be working in safer facilities be-cause BP will be making the necessary safety and health upgrades."

The settlement required BP to work closely with OSHA on any plans to restart the failed isom unit, to hire a process-safety consul-tant and to hire an expert on improving communications with

employees. The fines were imposed for five "egregious willful violations" of safety regulations, including failure to maintain its alarm
systems, and a host of other health and safety violations uncovered
in the six-month investigation.

By this time more than a thousand legal claims had been made
against BP, and attorneys were stacking up the lawsuits in the state
district court in Galveston County. Some members of victims' families had already been offered settlements, and some jumped on them
quickly, including Eva Rowe's brother, Jeremy.

Eva herself had spiraled into anger and despair in the long days
after losing her parents. The double funeral in Hornbeck was night-
marish for her. In between memories of fishing trips with her dad;
of listening to their favorite band, Lynyrd Skynyrd; and of family
vacations to California, the Bahamas, and Washington, D.C., she
kept thinking of her parents being crushed in a trailer or being
burned alive. Her older brother, his wife, and two children had come
in from Minnesota, and Jeremy seemed to be taking things much
more calmly. Eva was annoyed that attorneys had started showing
up at her door even before the funeral to offer their services, and
was bothered when she learned that Jeremy had already hired one.
After the funeral on Saturday and an Easter Sunday filled by death
rather than resurrection, an Easter she was supposed to have enjoyed
with James and Linda, Eva drove to her parents' trailer late that
night and found Jeremy and his wife loading up her parents' Ford
Explorer with duffel bags and furniture. She and Jeremy had agreed
earlier to wait before removing anything from the trailer, and now
it seemed he was hightailing it out of town with whatever he wanted
to take.[8] Eva lost it. She jumped out of her car, grabbed some duffel

bags, dragged them back into the trailer, ran back outside, slashed the tires on the Explorer, and then locked herself inside and cried uncontrollably.

The following months were no better. Eva reverted to the self-destructive behavior of her youth—drinking heavily, using drugs, eating and sleeping little, and generally venting her anger at everyone she came into contact with, even her longtime boyfriend. Her brother and other relatives testified against her when she went to court insisting that she be the sole administrator of her parents' estate, and the only friends who came around anymore seemed to want a piece of any future settlement they expected her to get from BP. Her boyfriend wanted to get married, but she told *Texas Monthly* she wasn't sure about his motives.

Finally a friend told Eva about Brent Coon, a Beaumont attorney who seemed to be a perfect match for her rough-and-tumble personality. Coon was a brash Texan in his early forties, a guitar player in his own band, Image 6, and a promoter of rock, country, and hip-hop concerts through his company Coondog Productions. Coon also had done some legal work for the Texas chapter of United Steelworkers of America, which had long been clamoring about safety issues at the BP plant. Eva Rowe's case presented a perfect opportunity to put those problems in the national spotlight and force changes. "My personal feelings were, from a number of conversations, that Eva would be very tough for BP to work with," Coon told *Texas Monthly*.

The more Coon dug into the case, and the more documents he was able to uncover through a court order, the more he and Eva Rowe became convinced they were dealing with a company that had put its employees at great risk solely in an effort to increase profits. There were the constant demands to cut budgets, the refusal to consider workers' concerns, the rewards for managers who produced higher profit margins. There were the callous e-mails, especially right after the tragedy, and the memos about calculating the costs of worker deaths against the benefits of reduced operating costs. There was the cost-benefit analysis of the "three little pigs." Eva Rowe grew more outraged with each damning detail that surfaced in a discovery process that yielded more than 6 million documents. She flatly concluded that BP had murdered her parents, and thirteen other workers alongside them.

A handful of Texas City survivors "reluctantly" agreed to meet in the fall with *Houston Chronicle* reporter Anne Belli to describe the horrors of March 23. More than half a year past the explosion, the memories were still vivid and searing and had taken a heavy emotional and physical toll. Kristof Harris, who had miraculously survived the blast after debris formed a protective cocoon around him, had actually organized a number of private gatherings for some of the "trailer survivors" to help them support one another. "It's almost like a healing get-together," he said. "You get to see everybody and realize that we're all going to live through this, we're all going to get on with our lives."[9]

"I like getting together with those people," agreed Ralph Dean, the forklift operator who helped pull his injured wife out of the rubble. Alisa Dean, then thirty-two, was doing her best to get past all the pain of losing her father that day and spending the next four months in the hospital recovering from many broken bones and severe burns. "You can't keep looking at all the bad," Alisa told the group meeting with Belli. "Because all you are going to do is be depressed." Even the news that she could no longer have children wouldn't break her spirits. "There are children out there that need parents," she said.

Jack Skufca, a BP employee who was injured in the blast, declined to meet with the group but sent some comments to Belli in an e-mail. "I remember everything about that day," he wrote. "I think about it every day and each night I lay sleepless in bed." He recalled being trapped in the rubble but unable to call for help. "I remember praying to God to keep me alive for my family's sake," he said. "I remember the pain that is indescribable and unimaginable."

Pat Nickerson, a BP engineer for twenty-eight years who watched the disaster unfold from his Jeep about 150 feet away from his office in the ill-fated trailer, also was haunted by the memories. "I remember standing there still looking in disbelief that there were so many innocent people dead," he said. Nickerson had returned to work for seven weeks not long after the blast, but his emotions proved too much to handle. He wasn't sure if he would ever return there again.

Ralph Dean had no hesitation on that subject. "That place was built on evil ground," he said, as his wife added, "They just need to level it and walk away."

A year after the accident, it was reported that a federal grand jury had been impaneled to consider possible criminal charges against BP. The Chemical Safety Board was deep into its investigation and BP's independent commission chaired by James A. Baker III was meeting regularly, interviewing employees and digging through documents to make an assessment of the company's safety programs.

OSHA was also busier than it had ever been in the previous fifteen years monitoring safety conditions for workers at BP refineries and others around the country. In April 2006, OSHA hit BP with one of the largest penalties ever assessed in the agency's history, fining the company $2.4 million for unsafe conditions at its Toledo, Ohio, refinery. Many of the citations, including thirty-two "willful egregious" violations, were for problems similar to those that led to the explosion in Texas City. "We found very serious problems, the same kind of problems that caused workers' deaths in Texas, and we want to make sure they're dealt with," said OSHA spokesman Brad Mitchell.

The first civil trial over claims for workers injured or killed in the blast, including Eva Rowe's parents, was set for early November 2006 in Galveston County. A few weeks before jury selection was scheduled to begin, Coon got an indication that federal prosecutors were making progress in their criminal probe of BP when he received a subpoena for depositions and documents he had compiled for his case.

Eva Rowe was more determined than ever to proceed with the trial. She agreed to an interview with Ed Bradley, the veteran correspondent with the CBS program *60 Minutes,* for a story to be aired before the trial. Rowe said that Bradley, who began his career covering the Vietnam War for CBS, had tears in his eyes as she talked about her parents' death and how much she missed them. It was to be Bradley's final interview before he succumbed to leukemia at age sixty-five on November 9, 2006. At the end of the story, Bradley asked Eva if she was prepared to go to the ropes with BP. "I'm ready," she said. "I'm ready to go to trial."

Eva's brother Jeremy had agreed to a settlement with BP for an undisclosed amount, but Eva was adamant that she wanted to face BP officials in court to demand answers about her parents' deaths. She knew it would be a very difficult trial. BP's attorneys had grilled her in a deposition questioning her drug use, her run-ins with police, her volatile personality. She was convinced BP had been spying on her, spotting cars with people holding cameras parked outside her house; BP's lawyers insisted in court that they did not conduct any surveillance on her. Coon had set her up with a bodyguard for protection twenty-four hours a day, but Eva still lived in fear for her life. None of that mattered. "I could have taken the money and shut up, but I couldn't let my parents' deaths be in vain," she said. "I had to expose BP for the bad people they are."[10]

But the pressure mounted as the trial date grew closer. Eva rejected numerous offers of substantial sums of money, determined that the company would not buy its way out of being held accountable in public. Coon told BP's top attorney, William Noble, that the only way Eva Rowe would consider a settlement is if the company agreed to full disclosure of the documents he planned to use in the trial.

On the night before jury selection was to begin, Coon and Noble made one last attempt to work things out over drinks at Garza's Kon Tiki Lounge on the bay in Galveston. Noble offered large donations to charities of Eva's choice, in addition to a settlement payment, but Coon kept pressing him on the issue of document disclosure. Finally Noble relented and Coon called Eva at 4:00 A.M. "We got what we wanted," he told her.[11]

Eva, staying at a rented beach house nearby, thought briefly about turning down the deal and sticking with her plan to face off against BP in court. But the opportunity to raise money for causes that would have been important to her parents prompted her to relent and accept BP's offer. "I was so proud that I accomplished something so great for my parents," she said.[12]

The agreement for BP to provide $32 million in charitable contributions would break down as follows: $12.5 million to the Truman G. Blocker Adult Burn Unit at the University of Texas Medical Branch at Galveston, where two dozen victims of the accident had been treated; $12.5 million to the Mary Kay O'Connor Process Safety Center at Texas A&M University, where refinery engineers study safety issues; $5 million to the College of the Mainland in Texas City, which offers courses in refinery and chemical plant safety; $1 million to the Hornbeck, Louisiana, school system, where Linda Rowe was a teacher's aide; and $1 million to St. Jude Children's Research Hospital, her parents' favorite charity. BP also agreed to establish a fund for victims of the explosion and would match any donations made to the fund up to $6 million.

Eva, of course, also received an undisclosed sum for herself, enough to ensure she would be taken care of for the rest of her life. And the BP documents that essentially indicted the company for negligence were posted on a Web site managed by Coon for all the world to see.

The settlement didn't soften Eva Rowe's views of BP in the least. When she appeared at a news conference in Texas City to announce the $5 million donation to the College of the Mainland, she was her usual blunt self. "I hate BP," she said. "I feel like they murdered my mother and father."[13]

While hundreds of other victims of the accident worked out

their own settlements with BP—not a single trial went all the way to a jury verdict before the company agreed to plaintiff demands— the investigations by the Chemical Safety Board, the Baker Commission and, most significantly, the U.S. Department of Justice, continued apace.

chapter four

SMART PIGS

BIG OIL HAS BEEN THE bully of the energy world since John D. Rockefeller's men ruthlessly stomped out Standard Oil competitors more than a century ago, and nowhere has that swagger been more evident in recent decades than in Alaska.

The riches of America's largest petroleum field on Prudhoe Bay made the oil companies the kings of the Last Frontier, and a laissez-faire regulatory climate developed in a state with some of the nation's most precious and pristine natural resources. Environmental values are strong in Alaska, to be sure, but when oil and gas taxes represent more than 75 percent of the state's revenues, the producers tend to wield the most political clout.

John Browne, who started his rise to the top of BP as an apprentice engineer in Alaska, acknowledged that the industry was in charge from the very beginning of the state's oil boom. "Limited regulation around industrial processes and health and safety, and a sense that we were opening up a frontier, gave us the feeling that we could write the rules," Browne said in his 2010 memoir, *Beyond Business*. "We made many mistakes at the operational level. Sometimes we were insensitive and did not recognize that what we were doing was wrong. Arrogance prevailed."[1]

The industry's power in Alaska today has been consolidated in three companies—BP, ExxonMobil, and ConocoPhillips—following the wave of oil mergers in the late 1990s and early 2000s. BP operates Prudhoe Bay and owns 26 percent of the assets, ExxonMobil and ConocoPhillips each have 36 percent, and a handful of small companies own 2 percent.

BP also controls 47 percent of the Trans-Alaska Pipeline System that carries crude eight hundred miles from the North Slope to the port of Valdez on Prince William Sound (ExxonMobil and ConocoPhillips split the rest). The pipeline, known as TAPS by industry insiders, is operated by the Alyeska Pipeline Service Company under the control of the three big companies. As the largest shareholder, BP has the greatest voice in Alyeska, and a BP executive is most often named as Alyeska's CEO. Some Alaska officials don't like to be reminded they are staring into the jaws of a shark when trying to regulate the industry. "I have had the most senior pipeline oversight officer in the state tell me flat out that BP doesn't own the Trans-Alaska Pipeline System," said Richard Fineberg, a journalist and consultant on the oil and gas business in Alaska for several decades. "That's how far we are into the belly of the beast."

And when the beast is angry, it attacks. The biggest prey by a long shot has been Charles Hamel, a former oil broker turned chief whistle-blower on environmental and safety issues that have plagued Alaska's oil industry since the first gusher was struck there in 1968.

A Connecticut native and a close family friend of the actress Sissy Spacek, Hamel started out as an avid booster of Alaska's burgeoning oil business in the 1970s when he became an aide to U.S. Senator Mike Gravel (D-Alaska), an old prep school roommate. "Among my duties as his assistant, I worked relentlessly to convince Alaska residents, commercial fishermen, natives, and the public that the oil industry would be good for Alaska and would surely build an environmentally sound pipeline and port terminal," Hamel told a House panel that was investigating industry operations in 1991. "Prior to construction, I traveled the 800-mile right-of-way from Prudhoe Bay to Valdez."

The experience led Hamel to become an independent broker, handling sales of oil and other commodities and arranging tanker shipments, and he acquired a few leases in Alaska in hopes of capitalizing on the oil boom. Life was good: He married a secretary to the late Senator Henry "Scoop" Jackson (D-Washington); he bought a Florida condo and a mountain ski lodge; and he took in as much as $100,000 a month from brokering deals.

But just as quickly as he found success, Hamel ran into trouble when he fought back against perceived abuses by the oil companies. First he had a run-in in 1979 with John Browne, then a BP executive roving between Alaska, Canada, and Cleveland, Ohio, where BP had taken a majority stake in Sohio. Hamel believed BP was ripping off one of his crude-oil clients on shipping fees, and he confronted Browne about it at a BP office in Cleveland. Browne and another BP manager refused to budge on the fees, and Hamel went away frustrated and angry.

Later that year, Hamel learned from one of his clients that the oil they had purchased through him arrived in a supertanker from Alaska heavily diluted with water. It turned out not to be an isolated incident. Over the next two years, more watered-down oil shipments were delivered to Hamel's customers, many of whom ended their dealings with him. By 1982, Hamel was running out of clients and his brokering business went into a tailspin.

From his days on Gravel's staff and later as a broker, Hamel had come to know many people in Alaska's oil industry, both in management and in the fields. Initially he was told by Exxon that the dilution problem was the result of malfeasance at a transfer operation at the Panama Canal, but he later learned from a friend inside the company that Exxon, BP, and other producers were intentionally shipping oil they knew was watered down. Workers at the Alyeska Pipeline Service Company also told Hamel that two sets of records were being kept on oil shipments—one made up of falsified lab tests showing acceptable amounts of water in the samples, and one with the actual amounts, often in violation of federal regulations.

Hamel filed a complaint with the Alaska Public Utilities Commission and Alyeska's senior laboratory technician, Erlene Blake, testified on his behalf that she had been ordered to falsify tests to enable diluted oil shipments to leave the port of Valdez. Alyeska denied her claims and demanded that she provide proof, which Blake was unable to do. After his complaint was dismissed, Hamel was told by another Alyeska employee, Bob Scott, that two of

Blake's supervisors had bragged they had broken into her locker and taken the doctored records before she could turn them over to authorities.

Putting Hamel out of business was probably the worst thing that Alyeska and its owners, BP and the other oil companies, could have done. Workers like Scott, embarrassed and disgusted by management's unethical tactics, began reporting other problems to Hamel, including intentional releases of hazardous air pollution at the terminal in Valdez, illegal toxic discharges into Prince William Sound, poor maintenance on the pipeline and related equipment, and inadequate response plans to deal with oil leaks and spills.

By the mid-1980s, Hamel said he had heard so many complaints that he concluded "the oil industry was turning Alaska into an environmental disaster. Employees I talked to in Valdez, friends I knew in the industry, people I had worked with for years, were all discussing the dismal performance of Alyeska in regards to their commitment to environmental and worker safety."

With his Washington experience, Hamel knew how to funnel the workers' concerns to the proper authorities at the U.S. Environmental Protection Agency (EPA) and other federal agencies. When the regulators were slow to respond, Hamel stepped it up a notch and sent information to reporters and to members of Congress. But no sooner had the House Interior Committee agreed to investigate in early 1989 when disaster struck just after midnight on March 24. The drunken captain of the *Exxon Valdez*, Joseph Hazelwood, slammed the oil tanker on to the reef in Prince William Sound, cutting a gash in the hull that eventually dumped about 11 million gallons of crude into the water. It would be one of the largest oil spills in history and one of the most damaging to the environment up to that point.

While the spill was Exxon's responsibility, the response was assigned to BP as the majority owner of Alyeska, and it was woefully unprepared. The Alaska Oil Spill Commission lamented that fact in its final report on the *Valdez* accident, although it didn't specifically cite BP. "With a well-prepared contingency plan, well implemented, the disaster of the *Exxon Valdez* could have been far less serious," the commission concluded. "Oil might never have reached shore."

Instead, the barge containing booms and oil-skimming equipment that should have been dispatched to the *Valdez* immediately was sitting unloaded, under repair, and under several feet of snow at an onshore facility, according to Dan Lawn, a former inspector for the Alaska Department of Environmental Conservation who was one of the first to arrive at the scene of the spill. Richard Fineberg, who went to the site as an observer for Alaska's governor three days after the accident, said he recalls flying over the spill in a helicopter and seeing few booms to contain the oil and few response vessels in the still-calm waters. Bad weather was expected soon, however, and at a dramatic meeting Fineberg attended late that night, the Coast Guard commander, Admiral Clyde Robbins, broke a pencil and threw it on the table in frustration after oil executives said they hoped the response barge would arrive the next day. What did arrive was a severe storm that began blowing the oil hundreds of miles along the coast and churning up the waters so that dispersants would no longer be useful.

The *Valdez* disaster did bring needed reforms to the oil business, but primarily on the shipping side, such as requirements for double hulls in supertankers and much greater response capability that must arrive quickly at a spill. Onshore problems such as pipeline corrosion and leaky valves had fallen off the radar during the spill's aftermath, and workers continued to see them go unaddressed.

A small but growing network of whistle-blowers kept information about the problems flowing to Hamel, who finally convinced the EPA and the General Accounting Office (GAO), the investigative arm of Congress, to get involved. BP and Exxon were hit with citations for air and water pollution at the *Valdez* terminal and were ordered to make costly improvements. The GAO issued a report blasting Alyeska and its owner companies for not doing enough to prevent pipeline corrosion, for not installing sufficient leak-detection systems, and for failing to prepare adequate response plans for spills.

Rather than moving aggressively to tackle the problems, though, Alyeska and BP, which had the biggest stake in the pipeline management firm, focused on trying to find out who was leaking information to government investigators. Alyeska, headed by a BP executive at the time, hired the Wackenhut security agency in 1990 to conduct an investigation, which would include spying on suspected whistle-blowers. Hamel was the primary target.

Wackenhut, a Florida firm that was a magnet for former FBI and CIA agents, put its director of special investigations, Wayne Black, in charge of the operation. His team watched Hamel's home in Alexandria, Virginia, videotaped his comings and goings, wiretapped his phone, stole his mail, and even sifted through his garbage looking for evidence of contacts with Alaska oil workers. Black also set up an elaborate scheme to trap Hamel the next time he traveled to Anchorage.

When Hamel went to one of his favorite spots there, Fletcher's Bar, an alluring blonde in a see-through blouse sidled up to him and struck up a conversation. Hamel was charmed, but avoided entanglement with her until a few days later on his return flight to Washington, when the woman was seated next to him on the plane. She told him she worked for an environmental group called EcoLit, which had a team of lawyers digging into problems in the Alaska oil industry. Perhaps the group could team up with Hamel, she suggested, inviting him to visit with her boss "Dr. Wayne Jenkins"— the alias used by Wackenhut's Wayne Black—in EcoLit's office conveniently located near Hamel's home in Virginia.

Hamel said later the woman caught him at a vulnerable time, when he was feeling overwhelmed by worker complaints from Alaska and frustrated by the difficulties getting authorities to respond. "I desperately wanted to go on with my life, to leave behind me the disillusionment that I felt, to do what other men at my age are doing— walking on the beach with their wife, enjoying the hard-earned fruits of their labor," Hamel told a congressional hearing in 1991. "Instead, the fruits of my labor were stolen from me, and the peace and contentment I tried to achieve were replaced by worrying and concern for those people who turned to me for help. Personally, these were terrible, dark nights for Kathy and me."

And so, with an offer of help from "EcoLit," Hamel took the bait. He wrote to his attorney in Anchorage, Julian Mason, that EcoLit was "the stuff that dreams are made of." The head of the group, "Dr. Jenkins," expressed outrage at the environmental crimes being committed by the oil industry and asked for Hamel's contacts, informants, and sources who could help expose the problems.

Hamel tried to be helpful without identifying any whistle-blowers, but combined with the information that Wackenhut had obtained from its spying on Hamel—including stealing his phone bills—the investigators were able to pinpoint many of his sources. Some of them were fired; one died from a heart attack after losing both his job and his home as a result of the probe.

"Obviously I did compromise many of them," Hamel said. "Inadvertently, of course, but nonetheless I let them down and I will always have to deal with that."

Hamel didn't learn he had been duped until nine months after Alyeska canceled its contract with Wackenhut, fearing that the investigators were going too far by trying to trap the chairman of the House Interior Committee, Representative George Miller (D-California), into accepting "stolen documents" from Hamel. A number of the Wackenhut investigators, ashamed by their gross invasions of Hamel's privacy, actually went to him in 1991 and confessed to the spying operation. Hamel filed suit in 1992 and received an undisclosed settlement the next year; the judge who oversaw the case, Stanley Sporkin, described Alyeska's tactics against Hamel as "reminiscent of Nazi Germany." The company ran full-page newspaper ads apologizing to Hamel, who became a national celebrity after the CBS program *60 Minutes* did a story on his case.

The harassment by Alyeska turned Hamel into an even greater hero to oil workers and managers concerned about the risks posed to them and Alaska's environment by shoddy industry practices.

In 1992, an inspector for Alyeska, Glen Plumlee, filed an affidavit citing numerous violations of quality-assurance standards for pipeline operations. The next year, Plumlee and six of his colleagues who felt Alyeska was ignoring their reports took their concerns to Congress and testified at a hearing on corrosion and other problems in the pipeline. Months after the hearing, a congressional investigator sent to Alaska uncovered a note written by a pipeline contractor that was an apparent blacklist of inspectors who had complained about operations, including Plumlee. "Former Alyeska Inspectors, Do Not Touch!" was written in the margin next to the names.

Most of the inspectors were terminated and had to fight for their jobs back using the Labor Department's whistle-blowers' protection, but their complaints to Congress did produce some results. The federal Bureau of Land Management (BLM) sent an audit team to Alaska to investigate how Alyeska handled reports of pipeline problems. The auditors discovered Alyeska employees had been manufacturing false reports and stuffing them in the files, but the ruse failed. The BLM issued a report citing deteriorating and corroding sections of pipe, faulty wiring and valves, and rampant falsification of inspection reports. Alyeska managers pledged to Congress that they would repair the lines and improve communications with workers.

If Alyeska learned any lessons from the experience, it apparently did not pass them along to its bosses at BP. In 1995, a conscientious worker for a BP contractor on the North Slope, Doyon Drilling, started keeping track of the times he was ordered to inject hazardous wastes such as used paint thinners into wells drilled on a man-made island northeast of Prudhoe Bay. Whenever the worker objected that the toxic dumping was illegal, he was harassed by other Doyon employees who continued the practice of mixing barrels of hazardous wastes with nontoxic drilling mud, usually during the long Alaska nights. Finally the worker, who has never been publicly identified, filed a complaint with authorities and an investigation began.

BP initially dismissed the complaint as a couple of isolated incidents at a remote field on Endicott Island. But later the company's own records showed the practice of dumping barrels of waste into wells had continued over a two-year period, sometimes with as many as thirty fifty-five-gallon drums unloaded at once at the site for disposal. BP pleaded guilty in 1999 to a felony count of violating hazardous waste laws, and the following year a federal judge sentenced the company to five years of probation and a $500,000 fine on top of the $6.5 million in civil penalties it had already paid, and ordered it to spend at least $15 million to improve environmental monitoring at all its drilling sites in Alaska and the Gulf of Mexico.

The guilty plea came not long after BP had acquired Amoco and ARCO, with its extensive holdings in Alaska, and CEO John Browne ordered across-the-board cuts of 25 percent in all of the expanded company's operations. The effects were noticed immediately by BP and Alyeska workers in Alaska. Six employees sent an anonymous letter to federal officials in 1999 arguing that neglect and maintenance cuts on the pipeline could lead to disaster. "It won't be a single gasket, or valve, or wire, or procedure, or person that will cause the catastrophe," the letter said. "It will be a combination of small, perhaps seemingly inconsequential events and conditions that will lead to the accident that we're all dreading and powerless to prevent."

Frustrated by the lack of response, dozens of BP workers pressed their case with Hamel. "We were concerned about BP's cost-cutting efforts undermining our ability to respond to emergencies, and

reducing the reliability of critical safety systems," said one of the letters workers sent to Hamel in April 2001. "We were concerned about
the lack of preventative maintenance on our equipment. We had
suffered a major fire, which burned a well pad module to the ground,
and nearly cost one of our operators his life." Hamel passed on the
concerns to Browne in a letter of his own. "Courageous 'Concerned
Individuals' contacted me for assistance in reaching you," Hamel
told Browne. "They have not succeeded in being heard in the past
two years in London, Juneau, or Washington. I am again a reluctant
conduit. They hope that you will take whatever action appropriate to
effect corrective action which would protect the environment, the
facilities, and their safety." Hamel sent a copy of the letter to President George W. Bush, who had just taken office a few months earlier with a promise to expand domestic energy development.

The following year a BP well exploded, killing one worker and
severely injuring another, prompting even more workers to become
whistle-blowers. One of them, Bill Herasymiuk, warned that a potential "catastrophe" could occur because of inadequate maintenance.
A number of them also testified about the problems at a congressional hearing in March 2002.

BP brought more problems on itself when it failed to promptly
report a 6,000-gallon pipeline spill in May 2003, an apparent violation of its probation in the toxic-dumping case. The spill was caused
by corrosion in one of BP's many pipelines on the North Slope that
carry crude directly from the well fields to collection stations and
processing facilities for transport to Valdez through the TAPS.
Shortly after learning about the spill, the commissioner of the Alaska
Department of Environmental Conservation (DEC), Ernesta Bal-

lard, wrote to the EPA saying BP was not getting enough oversight of its pipeline management, and asked the feds to step in and help. Ballard's letter went to Jeanne Pascal, the EPA counsel in Seattle who was monitoring BP's compliance with the terms of its probation.

Pascal consulted with Alaska DEC officials about pursuing criminal charges against BP for violating its probation, but the state regulators recommended against it and the EPA attorney agreed to back off. She continued hearing from workers sent to her by Hamel—at least thirty-five of them during BP's probationary period. Many were concerned about inadequate maintenance to prevent corrosion in the North Slope pipelines, some said BP was falsifying reports on corrosion, and all were terrified they would be fired if the company found out they had talked to the EPA, Pascal said years later.

In May 2004, Pascal and BP's probation officer at the federal court in Anchorage, Mary Frances Barnes, took the workers' concerns directly to BP's top managers and ordered them to do an internal investigation of maintenance programs on the North Slope. BP hired Vinson & Elkins, a Houston law firm that works closely with the energy industry, to conduct an audit. After reviewing documents and interviewing about forty-five employees, the firm concluded that BP was keeping its equipment in good shape and found no evidence that any reports had been doctored. "Nothing we learned in our investigation suggests that the field is, as a general matter, unsafe or prone to catastrophic failure," it said.

Vinson & Elkins did find one serious problem, though: The management style of Richard Woollam, BP's corrosion manager in Alaska. Woollam, the lawyers said, was "dictatorial" and had been dubbed by workers as "King Richard," a man who did not take criticism or complaints well. For instance, the audit report said, when a manager told him his employees were upset about staff reductions, Woollam responded, "Fire the whole bunch." The law firm recommended in October 2004 that Woollam be removed from management, and in January 2005 he was transferred to a nonsupervisory position at BP headquarters in Houston.

The calamity that North Slope workers had been predicting for years took place in early 2006 about 650 miles north of Anchorage on the frozen tundra near Prudhoe Bay. Just before 2:00 A.M. on Sunday, March 2, workers driving slowly past a section of thirty-four-inch pipeline covered by a caribou ramp smelled oil from their pickup truck and heard a bubbling sound when they stepped outside. After digging through snow and ice to uncover the pipe, they found a small hole with oil pouring out, and by the looks of the landscape around them, it appeared that the leak had been flowing for several days. Eventually response teams would find that more than 200,000 gallons of oil had spread over about two acres of snow and ice, the largest spill of petroleum on land in Alaska history.

It took very little time to determine that the leak had been caused by corrosion, and within days the federal Pipeline and Hazardous Materials Safety Administration issued orders requiring BP to repair any corrosion damage before restarting its pipelines, to develop better plans for reducing corrosion, and to review and improve its leak-detection system, which had apparently failed to sound alarms about the leak.

Cleanup crews braved temperatures as low as twenty degrees below zero to recover most of the oil over the next several weeks while the pipeline remained shut down, costing the state at least $1 million a day in oil revenues. Damage to the tundra was kept to a minimum by the frigid weather, since much of the oil congealed and was easier to scoop up. "Thank God this happened in the winter," commented Noah Matson, director of the federal lands program for the group Defenders of Wildlife, noting that few animals ventured into the region until warmer weather returned.

BP officials initially expressed surprise at the leak, insisting that the transit lines in the Prudhoe field where the leak occurred had been inspected 85 times for external corrosion and 139 times for internal corrosion since 1999. But the internal inspections were done from the outside with ultrasound devices that provide only rough images of the pipeline walls, the company acknowledged. Critics began to ask why BP hadn't used the more conventional "smart pigs"—electronic devices that crawl through pipelines and make accurate measurements of any areas where the walls have been thinned by corrosion. Smart pigs, however, are expensive to use and require temporary pipeline shutdowns. BP admitted that its transit lines on the North Slope had not been "smart pigged" since 1998. The company insisted it was using effective measures to inspect its lines, and was spending $71 million on the effort in 2006, up from $50 million in 2004. Congressional investigators would later uncover evidence that BP's former corrosion manager, Richard Woollam, had ordered cuts in smart pig operations to reduce costs under pressure from BP headquarters.

Fighting corrosion in the wet environment of Alaska is unquestionably a difficult task, particularly in areas where the pipes dip down to allow for wildlife crossings, such as the pipe that leaked in early

2006. And there are more than a thousand miles of pipelines criss-crossing the North Slope collecting oil for transport to Valdez in the TAPS. BP spokesman Daren Beaudo also maintained that as the Prudhoe Bay oil field was being drawn down, drillers were bringing up more viscous oil that can lead to more corrosion in the steel pipes. Maureen Johnson, senior vice president and manager of BP's Prudhoe Bay unit, also told reporters soon after the March spill that the company believed the corrosion that caused it had oc-curred within the past six to nine months, and would not have been detected if smart pigs had been run through the line a year or two earlier.

Nevertheless, the spill led the EPA to expand a criminal probe of BP that had been opened several months earlier, it was reported in April 2006, and almost simultaneously the Alaska Department of Environmental Conservation promised its own investigation and vowed there would be major penalties assessed against BP, with fines likely to reach into the millions of dollars.

Top Democrats on two House committees, though still in the minority in April 2006, launched their own inquiry and dispatched an investigator to Alaska to start collecting evidence. Near the end of April, a House Energy and Commerce subcommittee held a hear-ing on the pipeline problems and asked the head of the Pipeline and Hazardous Materials Safety Administration to testify. Stacey Ge-rard, the agency's acting administrator, revealed that not only had BP failed to use smart pigs to monitor for corrosion in its pipelines, the company also had stopped using "scraper pigs" that remove sludge that can cause corrosion. "It was our expectation that they would have been running those scraper pigs and that most companies do run the scraper pigs on a weekly to biweekly basis," Gerard said.

Representative Edward Markey (D-Massachusetts) asked Gerard why BP wasn't using the devices. "It can't be, bogusly, that BP doesn't have enough money," he said. "Or is it just another cost-saving measure, regardless of what the consequences are?"

"I can't speak to how much money BP has, but it certainly seems like they should have been running scraper pigs on a weekly basis so that this problem didn't build up and occur," Gerard said.

"And what is their explanation to you for their failure to run the pigs?" Markey asked. "Is it the same one I gave to my mother for not cleaning my room? I mean, what is the answer?"

"I think there is a question about how much of the deposits there are and questions about how to remove them," Gerard said. "We don't have a good explanation as to why they didn't start sooner, but obviously having not started sooner, the problem is more difficult today and is going to be hard to address, but we—"

"But what is their explanation for not doing their job?" Markey interrupted.

"We don't have an explanation for why they did not run scraper pigs," Gerard replied.

"No acceptable answer to you or no answer at all from them?"

"No acceptable answer."

BP spokesman Beaudo told reporters after the hearing that the company planned to increase its use of scraper pigs in the future. "We're going to improve on what we thought was already a very effective system—one that is supported by the fact that in the thirty-year history of operations, prior to the GC-2 transit line spill, no one can remember a leak occurring on a Prudhoe Bay crude oil transit line," he said.

Less than three months later in July, after BP reported earning $2.6 billion in profits from its Alaska operations in 2005, the company announced it had found another break caused by corrosion in one of its North Slope gas pipelines only a month after the March spill was discovered. Officials said they hadn't reported the leak earlier because only a small amount of natural gas had escaped before the leak was sealed.

Shortly after that, on July 21, BP ran a smart pig through three miles of North Slope lines to comply with the federal pipeline agency's orders, and "found 5,476 potential bad spots, including 176 places where corrosion might have chewed through 50 percent or more of the pipe wall," according to an *Anchorage Daily News* report several months later. Around the same time in July, BP said it was shutting down more than fifty wells in the Prudhoe Bay field because of leaks reported by oil workers. A small leak also was detected in another oil-transit line.[2]

With the oil operation he had helped build in the 1970s apparently on the verge of collapse, BP CEO John Browne rushed to Prudhoe Bay to try to assure Alaskan officials, Wall Street investors, and the American public that the system was being managed properly. "I believe that scrutiny will find that BP is a company committed to working to the highest standards in every detail," Browne said. "We've put in place a series of steps to ensure that that is the case and I take full personal responsibility for ensuring that those steps are carried through."

Three days later, on August 6, 2006, BP announced it was shutting down the entire Prudhoe Bay field, cutting off the flow of ap-
proximately 400,000 barrels of oil per day, or about 8 percent of U.S.
domestic production. Browne personally called Alaska Governor
Frank Murkowski to tell him the bad news, knowing the shutdown
would cost the state about $6.4 million a day in deferred income.

"We regret that it is necessary to take this action and we apologize to the nation and the State of Alaska for the adverse impacts it
will cause," said a statement issued by BP America Chairman and
President Bob Malone. "However, the discovery of this leak and the
unexpected results of this most recent smart pig run have called
into question the condition of the oil transit lines at Prudhoe Bay.
We will not resume operation of the field until we and government
regulators are satisfied that they can be operated safely and pose no
threat to the environment."

BP announced it would replace sixteen miles of its pipelines
from Prudhoe Bay, an operation that would take at least six weeks
and possibly several months. Oil prices on the world market began
shooting up immediately, and tempers rose proportionately among
members of Congress.

"It is appalling that BP let this critical pipeline deteriorate to the
point that a major production shutdown was necessary," said Representative John Dingell (D-Michigan) the top Democrat on the
House Energy and Commerce Committee.

Even one of the oil industry's staunchest supporters, Senator
Ted Stevens (R-Alaska), expressed outrage. "I am disturbed not only
by the fact that over the years, when I've taken members of
Congress up there—particularly senators and people from the

administration—we've been briefed that this is the safest area in the world, and how it's been maintained, and how they've got special procedures to check for corrosion and erosion and any sludge inside the pipeline," Stevens said. "As a matter of fact, it just wasn't done. And somehow or other, the regime for management failed to recognize it hadn't been done."

The shutdown showed "chronic neglect," said Representative Joe Barton (R-Texas), chairman of the House Energy and Commerce Committee, in an August 11 letter to Browne. Barton announced that his panel would hold a hearing on the shutdown as soon as Congress returned from its summer recess.

That hearing, on September 6, gave members an opportunity to excoriate BP for needlessly causing disruption in the oil supply when gasoline prices were soaring to nearly $3 per gallon.

"It seems to me that BP was betting the company and their field that this field would be depleted before major parts of the pipeline failed and needed to be replaced," Barton said. "Maybe it [BP] shouldn't operate the pipeline." Democratic lawmakers sniped at BP's claim that it was "Beyond Petroleum," suggesting that BP really stood for "Broken Pipelines" or "Bloated Profits."

BP America Chairman Malone did his best to offer an apology. "BP's operating failures are unacceptable," he told the committee. "They have fallen short of what the American people expect of BP and they have fallen short of what we expect of ourselves."

"I deeply regret this situation occurring on my watch," added Steve Marshall, president of BP Exploration in Alaska, promising that the company was moving to increase the pigging of its pipelines—much as its workers warned it should do years before.

Richard Woollam, the company's former corrosion manager, showed up at the hearing as requested by the committee, but immediately invoked his Fifth Amendment right not to testify and promptly left the room.

Charles Hamel, meanwhile, could not resist the opportunity to tell the world, "I told you so." On the eve of the House hearing, Hamel spoke at a luncheon at the National Press Club and hammered away at the oil company he had been prodding for decades.

"BP lies about everything," Hamel said. "They lie even when they don't have to lie." He predicted that spills would plague Prudhoe Bay for years. "It's like an old garden hose that leaks," he said. "And a patchwork won't fix it."

Operations at the Prudhoe Bay field were restarted on September 27, nearly two months after the shutdown. But investigations by the EPA and the U.S. Justice Department, along with a separate probe by Alaska authorities, were just moving into high gear.

chapter five

THE LORD'S DEMISE

THE CHRISTMAS HOLIDAYS IN 2006 were anything but cheerful for BP's CEO of eleven years, Lord John Browne of Madingley. Although the company's profits were soaring from steadily rising oil prices, environmental and management snafus in BP's Alaska operations and the still unfolding human tragedy in Texas had already helped push Browne to announce five months earlier in July that he would step aside at the end of 2008.

Browne had been summoned to a meeting with BP Chairman Peter Sutherland on Friday, July 21, 2006, and was told to put an end to speculation about his plans. The company had a mandatory retirement age of sixty that Browne would reach in February 2008, but the recent string of problems beginning with the Texas City disaster had raised questions about whether Browne would stay longer to resolve the issues from his watch or be forced to make an earlier exit. In addition, tensions had been building between the chairman and CEO since late 2005, when it was reported that Browne wanted to pursue a merger with the European oil giant Royal Dutch Shell that Sutherland opposed.

By most accounts, the forty-minute meeting with Sutherland was not a pleasant one for the proud member of the House of Lords. The physically slight Browne told friends afterward that he was taken aback by Sutherland's bullying manner. After mulling his future over the weekend, and following rampant speculation in London newspapers that brought even more pressure from Sutherland, Browne held a news conference on July 25, 2006, to announce that he would retire on December 31, 2008, a little more than ten months past his sixtieth birthday.

"A company isn't about one person," Browne said, repeating a line that Sutherland was reported to have used in his meeting with the CEO. He insisted he had a good relationship with Sutherland, but simply felt the time had come to hand the reins to another leader. "At the end of 2008 I will have been CEO of BP for over thirteen years and that is quite a long time," Browne said. "This has been a matter of discussion for the chairman, the board, and me for a very considerable time."

Browne could look back on his four decades at BP with some degree of pride. He had successfully rebranded the company as green and progressive with the Beyond Petroleum campaign—earning Browne the moniker "Sun King" in the British press—and he had separated BP from the rest of Big Oil by taking a bold stance for action on climate change. He had moved aggressively to expand with the Amoco and ARCO acquisitions, making BP the world's third-largest—and for a time the second-largest—energy company not controlled by a national government. And he had famously "flattened" BP's corporate structure, spreading responsibility among managers around the world and eliminating much of the top-down structure that gave the company a highly bureaucratic reputation for nearly a century.

"Once a conventionally hierarchical firm, BP is now organized in a flat way," wrote Howard Gardner in his 2004 book, *Changing Minds*. "Once an organization where responsibility was diffuse or altogether absent, it is now a company where each individual is expected to contribute directly to profits or to engage in creating or distributing knowledge that will ultimately increase profits. Those who cannot justify their contributions are rapidly and, some would say, ruthlessly dismissed from the company."[1]

The decentralization of BP also may have insulated the London headquarters from emerging issues at operations that stretched around the world. Browne himself had a tendency to surround himself with people he considered team players, a group he fondly called his "Ninja Turtles" after a popular cartoon. If directors and vice presidents in London were hearing reports of problems in any of the more than a hundred countries where BP had a presence, they might be reluctant to share them with the boss, preferring instead to build on the impression that all was well in Lord Browne's kingdom.

BP headquarters also may have helped disconnect itself from day-to-day affairs as managers at the local and regional level took on more responsibilities. The message from London always seemed oriented toward the bottom line, so the rank and file knew the best way to get ahead was to keep churning out reports of higher profit margins, no matter what the cost. If safety or environmental standards were lowered and risks increased, so be it, as long as the top brass was happy. This problem was evident in the years leading up to the Texas City accident. Refinery manager Don Parus and others preferred writing memos extolling recent profits, and downplaying the growing concerns in the workforce about safety. When he was asked at a September 2006 deposition if he had been aware of the threats to refinery workers before the March 2005 explosion, BP refining chief John Manzoni replied, "No. I think had I been aware that we could have had a catastrophic failure, we would have taken action earlier, different action." Instead there was a horrible disaster that cost the company billions of dollars.

The same principle of keeping problems in-house had guided the former BP corrosion manager in Alaska, Richard Woollam, who ran his division like a fiefdom and punished the serfs who dared to complain about the operations. It wasn't until an outside auditor recommended his removal that management realized they were missing critical communications from field workers and the pipeline infrastructure began to spring leaks. The result was not just a damaging environmental and safety record that shut down the nation's largest oil field at Prudhoe Bay, at great embarrassment to BP, but shareholders in the company were becoming increasingly intolerant of what some perceived to be negligence by top management.

The managers of one pension fund that held six thousand BP shares went to an Alaska court in October 2006 to make its indignation known to the company. UNITE HERE, a union representing industrial and service workers, filed a lawsuit in state court in Anchorage charging thirty-eight BP executives, including Browne and Sutherland, with actions that diminished the company's stature and significantly reduced the earnings of the union's retirement fund. The City of London joined the lawsuit two days later in what the *Evening Standard* described as "an unprecedented move in UK corporate history." By joining the case, London officials alleged that BP's managers had jeopardized the pension fund for thousands of police, teachers, and other public employees in the British capital. The lawsuit blamed BP's leaders for exposing the company to criminal sanctions and penalties that damaged its reputation and lowered its earnings.[2]

The plaintiffs cited not only the problems in Texas and Alaska as evidence for their charges, but also referenced another federal investigation that had become public the previous summer and further sullied BP's name. In June 2006, the Commodity Futures Trading Commission (CFTC) filed a complaint in U.S. District Court in Chicago alleging that BP traders had plotted to manipulate the propane market and illegally boost prices for a heating fuel used by millions of Americans, especially in rural areas of the Northeast. At the same time, the U.S. Justice Department filed criminal charges in Washington against a BP trader who admitted being part of a conspiracy and agreed to cooperate with federal investigators. BP denied it had engaged in market manipulation, but acknowledged it had fired seven employees "for failure to adhere to BP policies governing trading activities."

Propane, used in tanks to fire up barbecue grills and for heating fuel in about 7 million U.S. homes, is small potatoes for an energy behemoth like BP. But federal authorities monitoring trading volumes and taping conversations uncovered an alleged plan by BP employees to corner a piece of the $30 billion propane market in hopes of earning the company an estimated $20 million.

The schemers devised the plan in 2003 and launched it in 2004 "with the knowledge, advice, and consent of senior management," according to the CFTC complaint. A Houston manager of BP's gas trading, Mark Radley, was caught on tape describing the plan, saying, "What we stand to gain is not just that we'd make money out of it, but we would know from thereafter that we can control the market at will." A Chicago trader for BP, Dennis Abbott, was also charged and later pleaded guilty to one count of conspiracy.

Under the plan, BP traders purchased a large volume of propane in February 2004—nearly 90 percent of the available TET propane that goes mainly to heat homes and businesses in the Northeast—and temporarily kept it off the market. The result was a short-term jump of about 50 percent in TET propane prices, a significant increase for lower-income homeowners who rely on the fuel in winter. "They ultimately drove up prices by establishing a dominant and controlling position in physical propane while simultaneously withholding a certain amount from the market," CFTC enforcement chief Gregory Mocek told *The Wall Street Journal*. As a result, he said, "We believe many Americans paid higher heating bills in the winter of 2004."[3]

The government said it would seek financial penalties against BP, would try to recover damages for consumers, and could pursue criminal charges against BP executives. The scandal came just three years after BP had paid $2.5 million to settle a case with the New York Mercantile Exchange over alleged manipulation of crude oil prices. BP admitted no wrongdoing in that case, but vowed to strengthen its rules for traders and reinforce internal training and enforcement. "You can have all the rules you want in the world [but] it's the education of the individuals at the desks and the tone you set," Browne said in a 2004 interview with the *Journal* after that episode. "You have to keep reinforcing that again and again and again. The code of conduct is not a piece of paper dreamt up and put on the Web site. It is actually real. And if you find someone violating the code of conduct you fire them very quickly."

Three BP traders in Houston—Cameron Byers, Martin Marz, and James Summers—were suspended by the company a week after the federal complaint was filed. The news put renewed focus on trading regulations in Congress, where a number of key senators were already investigating price-fixing in energy markets. "More and more trading is being conducted by large oil and gas traders on electronic markets where there is no oversight," said Senator Carl Levin (D-Michigan), the top Democrat on the Senate investigations subcommittee. "It's time to put the cop back on the beat." Ultimately, however, a federal district judge in Texas dismissed all criminal charges against the individual defendants.

While the continuing federal probes into BP's energy trading, its safety programs, and its environmental management were weighing heavily on Lord Browne in early 2007, there was more pressure coming from shareholders concerned about the gold mine in BP's energy portfolio—its oil and gas program in the Gulf of Mexico. The record profits of $22 billion the company reported in 2006 were based largely on steadily rising oil prices, and as production was waning on Alaska's North Slope, BP needed new reserves to assure nervous investors about its future growth. The company's presumed savior in that effort was a gargantuan drilling rig called *Thunder Horse*, a floating platform capable of extracting petroleum from deep beneath the sea, where vast deposits containing billions of barrels of oil were believed to be encased. But in July 2005, *Thunder Horse*—weighing 60,000 tons and built over nearly a decade at a cost of more than $5 billion—was sidelined indefinitely by the forces of Mother Nature and the subsequent discovery of serious flaws in its construction.

BP had been drilling successfully in the Gulf's shallower waters since the 1980s, but the oil industry had known for years that the most promising reserves were likely to be found below waters a mile deep or more, and from there, beneath thousands of feet of salt and rock under the ocean floor. Several companies, including BP, spent much of the 1990s plumbing into Gulf waters a half mile deep or more, without much success. Drillers began referring to the Gulf as "the Dead Sea," sending an alarming message to corporate pencil pushers. BP headquarters even put a freeze on the budget for deepwater exploration for two years while geologists and engineers did a reassessment. Then, as the industry prepared to enter a new millennium with projections of fast-rising energy demands, oil and gas deposits in deeper waters became a critical mission, a new frontier. BP executives, led by Lord Browne, allocated up to $2.5 billion for the effort, about 20 percent of the company's total budget for exploration and production worldwide.

The U.S. government had encouraged the expansion with a 1995 law slashing royalty payments for deepwater leases in the Gulf, and Congress provided more incentives with a new energy law in 2005. By 2006 BP had acquired some 650 tracts in deep waters. It had also entered a partnership with ExxonMobil, which provided a 25 percent stake, to build the largest drilling platform in the world. Initially dubbed *Crazy Horse*, after the rock band that frequently played with Neil Young, the project was renamed *Thunder Horse* in 2002 after objections from the Lakota Sioux tribe that BP was disrespecting the name of one of its greatest chiefs. (Bob Malone, then chairman of BP America, personally apologized to tribal leaders and attended a peacemaking ceremony, after which he was named an honorary Lakota.)

Thunder Horse was a genuine technological wonder of the energy world. Standing 420 feet above the surface when semisubmerged in the water, and stretching the length of three football fields, the rig was capable of drilling up to 30,000 feet below the sea floor, with the capacity to produce as much as 250,000 barrels of oil and 200 million cubic feet of natural gas on a daily basis. It was positioned about 150 miles southeast of New Orleans above a field believed to contain at least 1.5 billion barrels of oil. Three successful discovery wells had been drilled in the field from 1999 to 2001 at depths of between 25,000 and 30,000 feet, beneath more than a mile of water. Production from the *Thunder Horse* platform was scheduled to begin in 2005.

But in July of that year, after the well field on the bottom of the Gulf was nearly ready to go into production, along came Hurricane Dennis. Workers evacuated the rig, fully confident that the mammoth structure built to withstand the worst of storms would be fine. But the day after Dennis blew through, word came from a ship that passed the platform: It was listing at a thirty-degree angle like a crippled ocean liner on the verge of sinking.

A BP crew that arrived two days later discovered that *Thunder Horse* had not actually been damaged by the storm. Instead, a section of pipe with a valve installed improperly allowed water to flow freely through the ballast tank system, setting off a chain reaction of flooding that tipped the rig out of balance. Days of pumping enabled crews to right the platform, but it would require nearly another year of cleanup work, including restoration of dozens of huge pumps that had been damaged, before the rig would be ready for drilling again. Then BP needed to test the system of pipes and other equipment that had sat idle on the ocean floor for months while the platform was being repaired. Pressure tests on the well field conducted in May 2006 turned up a far more serious problem: some of the supposedly superstrength welds in the pipes had developed leaks, probably due to the intense pressures and corroding influences of the deep, dark sea. In effect, Hurricane Dennis had done BP a favor. Had the company been able to start production on schedule in 2005, the hot petroleum surging through the pipes under high pressure, combined with the effects the sea was having on the welds, might have eventually led to a catastrophic blowout the likes of which had never been experienced in the Gulf.

Now BP was faced with having to completely rebuild its sea-bottom well field, including redoing the welds to ensure they would last for years under the extreme conditions. That meant dismantling the entire field, sealing wellheads, bringing up pipes and equipment to make new welds, and adding coatings of anticorrosive materials. The work would further delay the start of production at *Thunder Horse* until at least 2008, BP concluded.[4]

No one ever said the technological challenges of deep-sea oil production would be easy. "This is as close as we get to the space age on earth," BP's vice president for production in the Gulf, Kenny Lang, told *The New York Times* in November 2006.[5] There were bright spots in BP's well-financed Gulf operations. In August 2006, BP announced the discovery of an 800-foot deposit of oil-bearing sands six miles down about 250 miles southwest of New Orleans. It was one of a dozen major discoveries by oil companies in the deep waters of the Gulf since 2001. Another potential gusher not far from BP's August discovery was announced by Chevron and two partners the following month; it was a field estimated to contain anywhere from 3 billion to 15 billion barrels of oil, enough to add as much as 50 percent to the known U.S. reserves at that time of 29 billion barrels. The potential for a boom in deepwater drilling was getting stronger and stronger, but so was the competitive pressure to produce.

Lord Browne was carrying all these issues into the new year in 2007 when he was smacked by yet another challenge, this time a deeply personal one. A Canadian man, Jeff Chevalier, who had had a four-year relationship with Browne that was broken off in 2006, had recently sent the BP CEO an e-mail. "I do not want to embarrass you in any way but I am becoming concerned by your lack of response to my myriad attempts at communication," the twenty-six-year-old Chevalier wrote to the fifty-eight-year-old Browne. Then, in early January, Browne received a call from a London tabloid saying it planned to out him in a story with details provided by Chevalier. Browne said later that the prospect of a tawdry story about his homosexuality, something he had zealously tried to keep private all his life, horrified him. He contacted his lawyers, who quickly obtained an injunction blocking publication of the story. A week later Browne announced that his retirement date from BP would be moved up to July, just six months away. Browne clearly knew that a rough road lay ahead, and he wanted to be sure the company was prepared to name his replacement.

The long, winding chain of dominoes that made up Lord Browne's career at BP then began to fall.

First came the report of the group led by James A. Baker III that had spent more than a year evaluating BP's safety programs. The commission issued a scathing indictment of company management on January 16, 2007. "Based on its review, the panel believes that BP has not provided effective process safety leadership and has not adequately established process safety as a core value across all its five U.S. refineries," its report said.

Instead, the panel found, BP managers had focused on the personal safety of its workers—preventing slips and falls, wearing hard hats and safety belts—at the expense of making sure the complex process of refining oil was conducted properly at all stages. "BP mistakenly interpreted improving personal injury rates as an indication of acceptable process safety performance at its U.S. refineries," the commission concluded. "BP's reliance on this data, combined with an inadequate process safety understanding, created a false sense of confidence that BP was properly addressing process safety risks."

BP managers failed to keep track of incidents or near accidents at all its refineries so lessons could be learned from them, and did not provide adequate resources to maintain a high level of safety, the panel said. "It is a very significant finding that BP does not effectively investigate incidents throughout the corporation," Daniel

Horowitz, a spokesman for the commission, told reporters. "If you're not learning from near misses, you're not in a position to prevent major disasters like the one in Texas City."

The failures were compounded by a heavy turnover of plant managers, a stressed workforce maintaining long shifts for many consecutive days, and poor lines of communication between supervisors and front-line employees. Not only in Texas City, but at its refineries in Toledo, Ohio, and Whiting, Indiana, BP had "not established a positive, trusting, and open environment," the report said.

The Baker Commission recommended fundamental changes in the safety culture at BP, and the appointment of an independent monitor to track the company's progress over the next five years. CEO Browne said an overhaul was already under way. "Many of the panel's recommendations are consistent with the findings of our own internal reviews," he said. "As a result, we have been in action on many of their recommendations for a year or more. Our progress has been encouraging but there is much more to do."

Browne put it more succinctly in a conference call with reporters. "BP gets it, and I get it too," he said. "I recognize the need for improvement." He insisted, though, that the company never put profits over safety.

Two months later, the federal board that had investigated the Texas City accident for nearly two full years drew a different conclusion. "Cost-cutting and failure to invest in the 1990s by Amoco and then BP left the Texas City refinery vulnerable to a catastrophe," the Chemical Safety Board said at the top of a 341-page report on the most exhaustive examination in its nine-year history.

The CSB concurred with the Baker Commission that BP's safety culture was flawed by focusing too much on personal safety and failing to track problems in refining processes. But it went far beyond Baker's panel in blaming company managers for not ensuring that an appropriate level of investment was made in the workforce and its equipment in Texas City. The isomerization unit that exploded, killing 15 workers and injuring 180, had alarms that did not function, equipment for handling overflow gases that was badly outdated, and a staff that was fatigued and stretched too thin, the board said.

The accident followed companywide budget cuts of 25 percent ordered by BP in 1999 and another 25 percent cut in 2004, on top of deep cuts made by Amoco in the decade before BP acquired the company in 1998, the board said.

"The Texas City disaster was caused by organizational and safety deficiencies at all levels of the BP Corporation," it concluded. "Warning signs of a possible disaster were present for several years, but company officials did not intervene effectively to prevent it."

BP issued a statement reiterating that it took full responsibility for the accident and was making safety improvements as recommended in January by the Baker Commission. But BP spokesman Ronnie Chappell was quoted in the statement saying BP had "strong disagreement with some of the content of the CSB report, particularly many of the findings and conclusions." He was not specific, but the company's consistent line since the tragedy had been that budget cuts were not a factor in the accident.

The CSB also sharply criticized a sister federal agency, the Occupational Safety and Health Administration (OSHA), for failing to monitor safety and enforce regulations at refineries across the country. "OSHA's national focus on inspecting facilities with high personnel injury rates, while important, has resulted in reduced attention to preventing less frequent, but catastrophic, process safety incidents such as the one at Texas City," the board said. There were ten fatalities at the Texas City refinery in the previous twenty years, including three in 2004 when not a single OSHA inspection was conducted. In the same twenty-year period, OSHA charged BP with only eighty-five willful or serious citations, and collected only $77,860 in fines, the CSB noted.

But the greatest criticism was reserved for BP management. CSB Chairwoman Carolyn Merritt was especially blunt. "The combination of cost-cutting, production pressures, and failure to invest caused a progressive deterioration of safety at the refinery," she said. Merritt said the BP board of directors shared responsibility by demonstrating "ineffective or nonexistent" oversight of the company's safety culture.

Merritt took her message to the refinery industry as a whole on the same day her board's report was issued. "Somebody has to be asking the question: 'What is happening, and is this being done?'" she said in a speech at the National Petrochemical & Refiners Association annual meeting in San Antonio. Her speech served as a warning that if refiners didn't do more to protect their workers, a regulatory crackdown would surely follow. "Performance is what's going to prevent overregulation," she said.[6]

The safety board leader repeated the refrain in May 2007 at a congressional hearing on BP's pipeline problems in Alaska. There was a "striking similarity" between the factors that led to pipeline and equipment leaks that caused the shutdown of the Prudhoe Bay oil field the previous year and the causes of the Texas City accident, Merritt told the House Energy and Commerce Subcommittee on Oversight and Investigations. "Virtually all of the seven root causes identified for the Prudhoe Bay incidents have strong echoes in Texas City," she said, citing BP's cost-cutting, production pressures, and poor communications with workers.

Merritt said she had reviewed an audit of the Prudhoe Bay problems prepared by Booz Allen Hamilton under contract with BP, and it concluded that the company's budget cuts in Alaska were "largely driven by top-down targets" that were "considered sacrosanct" by company headquarters. The cost pressures led to staff reductions, particularly in the corrosion-control program, and to "the deferral of integrity projects," she said.

"The CSB report [on Texas City] and the Booz Allen report [on Alaska] point to similar cultural factors within BP, in both its upstream production and downstream refining operations," Merritt concluded. "The similarity in the two reports underscores how safety culture truly is set at the top of a corporation. After all, the upstream and downstream sides of BP have separate reporting lines all the way to the Group Chief Executive and the board of directors in London."

Evidence uncovered by the House subcommittee, including e-mails from BP managers in Alaska, seemed to confirm that budget cuts in pipeline maintenance contributed to corrosion problems that produced the leaks. "These documents show that cost-cutting pressures on Prudhoe Bay operations were severe enough that some BP field managers were considering reducing or halting the range of actions related to preventing or reducing corrosion," said Oversight and Investigations Subcommittee Chairman Bart Stupak (D-Michigan). As an example, Stupak read from a 2001 e-mail written by Richard Woollam, the former BP corrosion manager who had refused to testify at a House hearing on the Prudhoe Bay shutdown in September 2006:

"As you know," Woollam wrote to the staff in the anticorrosion division, "we are under huge budget pressure for the last quarter of the year and therefore we have to take some disagreeable measures." Among those measures he listed shutting down certain "inhibition systems" for reducing corrosion and discontinuing the use of some chemicals that help prevent corrosion. "These need to happen as soon as possible," Woollam wrote.

Stupak noted that maintenance cuts in Alaska between 1999 and 2006 came during a period when BP earned more than $106 billion in after-tax profits. "As a result of BP's poor management of Prudhoe Bay," he said, "the public are the ones who ultimately are left footing the bill as the costs of supply interruptions are passed on to them in the form of higher prices at the pump."

Bob Malone, who had been named chairman and president of BP America after the Alaska problems surfaced in 2006, told the subcommittee that BP was moving swiftly to reform. "Today, I want to assure you that we get it," Malone said. "We have learned the lessons of the past." He also went further than any other BP executives in acknowledging the effects of budget cuts on safety. "It is disturbing to me if even one person in our organization thought of options of placing budget considerations over the safety and the integrity of our operations," Malone said. "It is clear that budget impacted our culture and that we stopped being curious."

John Browne was not part of the hearing, though a few weeks before he had made a dramatic public appearance that was to be his last as BP's CEO, with a speech aimed at demonstrating that BP's leadership still had a strong social conscience. On April 26, 2007, Browne returned to Stanford University for the ten-year anniversary of his groundbreaking address on climate change, and repeated that it was still a major concern.

"Now, the consensus of informed opinion is that the risk is both greater and closer than was imagined then," he said. "Ten years of great scientific work have improved our knowledge of many aspects of the issue." His lengthy speech went on to outline the scope of the climate change problem and steps BP was taking to address it, and he lamented that others had not followed the company's example with action rather than rhetoric. "This is not just an issue of branding or public relations, though it is fascinating and indeed very encouraging that so many companies feel they should be advertising in this way," he said. But, he added, "A lot of words have been spoken but only very limited actions have been taken." Browne pledged that the company would continue to be a world leader on the issue. "My successor as CEO of BP and the whole of BP's team intend to continue to play their part in leading this change in a practical way," he said. "And I hope to do the same, in a relevant role, even after I have stepped down as CEO of BP."

Just five days later, on May 1, 2007, Browne was forced to resign. The four-month court battle over publication of the story about Browne's affair with Jeff Chevalier ended badly—BP Chairman Sutherland would call it a tragedy.

Browne had attempted to use Britain's strict libel laws to block the story, and under the circumstances, with an apparent blackmail attempt at work, he seemed to have an excellent chance of succeeding. But Browne's pride led him to make a critical error in judgment during the trial. He testified that he had met Chevalier in 2002 while jogging in a London park. Chevalier, however, had provided evidence that Browne linked up with him through an online escort service. The judge, angry that Browne had lied to the High Court, dismissed the case, adding that since Chevalier alleged that Browne had shared information about BP with him during their relationship, BP's shareholders and board had a right to know how the company's resources had been used. Browne took his case to the British Court of Appeal, but it rejected his arguments and ruled that Associated Newspapers, publishers of the *Daily Mail* and the *Evening Standard*, could run stories about Chevalier's assertions. Browne again appealed, this time to his peers in the House of Lords. That final appeal was also denied on Tuesday, May 1.

Browne quickly issued a statement to announce his resignation. "For the past forty-one years of my career at BP, I have kept my private life separate from my business life," he said. "I have always regarded my sexuality as a personal matter, to be kept private. It is a matter of personal disappointment that a newspaper group has now decided that allegations about my personal life should be made public."

Browne admitted that he had a four-year relationship with Chevalier, but insisted there was no corporate misconduct. "The allegations are full of misleading and erroneous claims," Browne said. "I deny categorically any allegations of improper conduct relating to BP. The company has confirmed today that it has found no such wrongdoing." But Browne did acknowledge making false statements to the court. "My initial witness statements, however, contained an untruthful account about how I first met Jeff," he said. "This account, prompted by my embarrassment and shock at the revelations, is a matter of deep regret. It was retracted and corrected. I have apologized unreservedly, and do so again today.

"These allegations will result inevitably in considerable media attention for both myself and BP," he concluded. "What matters to me is BP, what we have achieved during my tenure, and the bright future ahead."

Sutherland said the company had done an extensive review of Chevalier's charges and found no basis for them. "That review concluded that the allegations of misuse of company assets and resources were unfounded or insubstantive," the BP chairman said in a statement. "The Board of BP has accepted John's resignation with the deepest regret," he said. "For a chief executive who has made such an enormous contribution to this great company, it is a tragedy that he should be compelled by his sense of honor to resign in these painful circumstances." Browne's departure was painful for him in other ways. BP said he would lose a bonus worth more than $7 million and would be dropped from the company's performance share plan for 2007–2009, estimated to be worth up to $24 million.

BP immediately appointed Tony Hayward, just three weeks shy of his fiftieth birthday, as Browne's successor. Hayward, a geologist who had been Browne's executive assistant and later headed BP's exploration and production operations, had already been tapped to replace Browne after his anticipated retirement in July. Before he was chosen over at least four other leading candidates for the position, Hayward had seemed to be distancing himself from Browne in the previous year. It may have helped push him to the top of the field. "We have a management style that has made a virtue out of doing more for less," he said in a memo that was leaked in mid-2006. "The mantra of more for less says that we can get 100 percent of the task completed with 90 percent of the resources—which in some senses is okay and might work, but it needs to be deployed with great judgment and wisdom. When it isn't, you run into trouble."

As Hayward set about to change BP's culture and return to a more deliberate, nuts-and-bolts approach to business, environmental and safety issues continued to surface.

The *Chicago Tribune* reported in July that BP's refinery in Whiting, Indiana, just southeast of Chicago on the shore of Lake Michigan, had received an exemption from strict new limits on releasing mercury into the lake, the source of Chicago's water supply.[7] The toxic metal that can cause brain damage in tiny doses was already a growing health concern in the Midwest, with scores of coal-fired power plants spewing mercury into the air and water and prompting warnings about eating too much fish from the Great Lakes. Now BP was being told by Indiana officials it could have another five years to cut its mercury discharges into Lake Michigan, and would be likely to get a further extension if needed. BP was in the midst of expanding the Whiting refinery to process higher crude volumes expected from the Canadian oil sands, and a controversy was raging about Indiana's plan to raise the allowable limits on ammonia and other pollutants in BP's water permit. After a huge public outcry joined by a host of members of Congress, BP backed off its request to increase its discharges in August, but environmental groups vowed to closely monitor every stage of BP's refinery expansion.

In Texas, thousands of claims against BP stemming from the March 2005 refinery explosion continued to work their way through the legal system. The endless litigation was almost daily fodder for *The Houston Chronicle,* and some of the more dramatic developments inevitably were picked up by the national media, keeping the accident fresh in the public consciousness.

Four workers who were injured in the blast settled their case in September 2007, but not before making appearances in court to describe the impacts. Their attorney, Brent Coon, who had won the settlement for Eva Rowe in 2006, summed up the workers' feelings: "They have emotional scars worse than any physical scars," Coon told the Associated Press after the settlement was announced. "When you see your buddies, co-workers, friends disintegrate before you, the psychological impact of being thrown into that kind of inferno— hell on earth—no one would understand."[8]

Then, in October, the full force of the U.S. Justice Department came down on BP, and the deadly accident in Texas, the oil spills in Alaska, and the scandal over price-fixing of the propane market were back in the national news all at once. Acting Attorney General Peter Keisler—an interim replacement for Alberto Gonzales, who stepped down as attorney general in September amid a number of swirling controversies—announced that BP had agreed to pay $373 million to settle criminal and civil charges stemming from three major federal probes.

The largest part of the proposed settlement would end criminal prosecution against BP for manipulating the propane market. Under the agreement, pending acceptance by a federal judge, BP would pay a penalty of $100 million, fines of $25 million to the U.S. Postal Service and $125 million to the Commodity Futures Trading Commission (CFTC), and restitution of $53 million. The acting chairman of the CFTC, Walt Lukken, said the size of the punishment reflected "outrage that these people were taking advantage of innocent consumers."

The Texas case, charging that BP failed to meet clean air standards in allowing harmful chemical vapors to be released at its Texas City refinery on March 23, 2005, would cost BP $50 million in return for pleading guilty to a felony. Officials said it would be the largest penalty ever assessed for a single violation of the Clean Air Act.

As punishment for its March 2006 spill of 206,000 gallons of crude oil on the Alaska tundra, BP would pay $20 million and plead guilty to a misdemeanor violation of the Clean Water Act. The penalty would include a $12 million fine, $4 million in restitution to the state of Alaska, and $4 million for research by the National Fish and Wildlife Foundation.

"The actions that we are taking today reflect that there were some very serious problems at the company," said Keisler, the acting attorney general.

BP America Chairman and President Bob Malone confirmed the company would agree to the settlement if it was accepted by the federal court, and he issued another statement of regrets. "These agreements are an admission that, in these instances, our operations failed to meet our own standards and the requirements of the law," he said. "For that, we apologize."

The company obviously wanted the settlement to put the trage-
dies and embarrassments of the past thirty-one months in the past,
once and for all, but it would soon learn that such hopes were premature. Attorneys for victims of the Texas City accident howled
that the settlement in the Texas case was "shockingly lenient" and
woefully inadequate in light of BP's "terrible misconduct." They
argued the penalty should be closer to $1 billion, approximately the
amount of profits the company made from the Texas refinery in the
year before the accident. BP defended the settlement amount, arguing that it was spending $1 billion to upgrade the refinery and had
already paid $1.6 billion to compensate victims of the accident.

In the Alaska probe, a former investigator for the Environmental
Protection Agency who worked on the case against BP came forward to call the $20 million settlement, with only a misdemeanor
plea, a sellout by federal prosecutors. "Everybody was convinced we
had a humdinger of a case," Scott West told *Newsweek* magazine.
"There was a corporate philosophy that it was cheaper to operate to
failure and then deal with the problem later rather than do preventive maintenance." West and another retired EPA investigator, Bob
Wojnicz, said they were then were ordered to back off. "We were
told, 'main Justice wants this wrapped up,'" West told *Newsweek*.
He and Wojnicz also claimed that BP's legal team had used their
Washington connections in the Bush administration to get a blanket settlement for all three pending investigations.[9]

BP's settlements of the propane trading charges and the Alaska
investigation were accepted by federal judges overseeing those cases,
but attorneys for plaintiffs in Texas formally challenged the terms
of the Texas City deal, leaving it to a judge to sort out all the issues,

as soon as one could be found. The first two judges assigned to the case disqualified themselves because of past connections to BP or some of its expert witnesses in the state court cases. Finally, U.S. District Judge Lee Rosenthal in Houston took over the case and said she would begin hearing testimony from victims of the accident and their families in early 2008.

If Tony Hayward wanted to move past the troubles that developed under his predecessor, Lord Browne, he would have to wait a while longer.

chapter six

LIARS VERSUS POLITICIANS

THE TRIAL OF NEODESHA V. *British Petroleum* began on August 23, 2007, three years and five months after the lawsuit was filed charging the company with fraud and negligence in its handling of oil-refinery contamination in the city. Legal teams for both sides spent the pretrial period taking depositions, conducting studies, digging through documents, and filing scores of motions, but there was little effort by BP's attorneys to seek a settlement before the case went to trial. The fact that there were no big media outlets in rural Kansas may have played a role—BP apparently felt it could slug it out in complex environmental litigation without sustaining any damage to its carefully crafted green image.

The setting for the David versus Goliath battle was a small brick courthouse in Erie, Kansas, about twenty-five miles east of Neodesha. Presiding over the trial was Allen County District Judge Daniel D. Creitz, rated by the Kansas Commission on Judicial Performance as one of the fairest judges in the state. A commission study found that "100 percent of the attorneys surveyed and 74 percent of non-attorneys perceived Judge Creitz to be completely neutral in criminal cases."

The attorneys had warned the jury pool going into the case that the trial was expected to last three months or longer, and the result was that half the jurors selected were over the age of sixty, since few working people could afford to take that much time away from their jobs. The oldest was an eighty-five-year-old woman with thinning hair dyed red; the youngest was a thirty-five-year-old, prematurely gray man who wore crisp dress shirts and took extensive notes with his left hand.

The city's case was presented by John M. Edgar, a distinguished, graying Texan in his midsixties who had settled in Kansas City to build a small law firm with his two sons. BP's team of lawyers was more than twice that size, with five women and four men led by Steven Lamb, a native Texan in his forties who was bald and heavyset, and who had learned how to handle corporate environmental cases in the Los Angeles office of a top national litigation firm, Zelle Hofmann.

In his opening arguments for the plaintiffs, Edgar pointed out that BP acknowledged it was responsible for the pollution it had inherited in 1998 when it bought Amoco, which had run the refinery in Neodesha until it was shut down in 1970 after seventy-three years of largely unregulated operation. BP even agreed that the contamination should be cleaned up if it posed a threat to health and safety, Edgar said. The plaintiffs were not alleging any health damages in this case—those claims could be made later by victims of the pollution if BP was found to be at fault, he noted. The community in this case was only arguing that BP had flat-out lied about the extent of the contamination, about the threats that it posed, and about their intention to address any problems that were found.

"It was a huge liability," Edgar told the jury. "And they were focused on trying to manage, limit, and terminate that liability." The company had a two-pronged strategy, he said: Control the state agency that was overseeing the site, the Kansas Department of Health and Environment (KDHE), and make sure the community didn't file a lawsuit demanding a full and costly cleanup. "They knew that the state, particularly the KDHE, was understaffed, underfunded, and underqualified compared to what the BP resources and scientists were," Edgar argued. And in the community, he said, BP orchestrated a huge public relations campaign that included door-to-door visits, a PR team that was trained to stay on message, surveys, advertisements, meetings with officials and local groups, handouts, an office in town—all aimed at convincing Neodesha that the pollution was under control.

"They had four messages, and they were four lies," Edgar said. "One, we're cleaning it up; two, the contamination is stable; three, there is no risk; and four, we know where it is."

The next day, Lamb had his turn to lay out BP's defense for the jury. "We don't think this is a case about cleaning up the mess," he said. "We think it has very little to do with cleaning up the mess. We think it has a lot to do with politicians and it has a lot to do with—I'm sure you are surprised—money." When he said "politicians," Lamb practically spat out the word, and sneered as he did so at the city officials from Neodesha sitting on one of the benches in the crowded courtroom. The plaintiffs, he said, wanted $650 million in damages plus a remediation project that would cost hundreds of millions more.

Such an extensive cleanup effort was not warranted, Lamb said, citing studies from the KDHE and an engineering firm hired by BP to assess the refinery site. Air tests had been conducted all around the site and no vapors were found that could pose a hazard, he said. Contaminated groundwater beneath the site was not being used by anyone because Neodesha's water supplies came from the Verdigris River running through the city. The aquifer itself was beneath three to five feet of clay soil, a barrier that kept any of the pollution from reaching the surface. "There is no pathway" for human exposure to the pollution, Lamb said.

"We had refinery contamination," he said. "We absolutely admit that. There's no question about it. That's not the issue. The issue is what is the nature and extent of the contamination, and what are the damages?"

BP was not just sitting on a toxic waste dump and doing nothing, Lamb contended. A corrective action study was approved by the KDHE in 2005 and extensive testing and some remediation work was continuing, with BP contractors actively pumping out contamination and treating it, he said. "We believe today there's no risk," he said. "If we thought there was a risk we would do something about it. If KDHE comes in and says, 'Hey, there's a risk, do this, do that,' we're going to do it."

Finally, Lamb argued, if property values in Neodesha were damaged by the pollution, why hadn't Wilson County reduced any of the property assessments in the city?

Edgar and his son, David, then proceeded to lay out the plaintiffs' case with a parade of witnesses that would stretch over three months.

The first was mayor and town banker Casey Lair, who testified that initially he accepted BP's arguments that the pollution wasn't a problem. "I believed them," he said. "They were the experts. I mean, this is the second-largest oil company in the world. They told me something, everybody in the place believed that. They were in our neighborhood, they were cleaning it up."

Lair learned later through studies conducted by the city that he, his wife, and his two children were living above a pool of arsenic. Their house was about a half mile from the old refinery site, but apparently the contaminated plume had been gradually flowing underground in their direction over the years.

City Administrator J. D. Cox testified that he, too, initially felt reassured that BP was addressing any problems at the refinery site. "There was no need for us to take action," he said. "I felt relieved to know that, well, they have identified it, at least it's the right thing to do and, yes, we relied on them."

Jim Scheussler, a city commissioner and management employee at the Cobalt Boats factory on the old refinery site, described how BP had asked him to serve on a community advisory board to recommend steps for addressing the pollution in 2003. Scheussler said he went to the meetings with an open mind, but quickly understood that throughout the process, BP made the rules, BP provided the information, and BP drafted the "cleanup plan" that was eventually recommended—a process of "natural attenuation" in which Mother Nature would remove the pollution herself over time, though it could take from five hundred to a thousand years.

Scheussler said he discovered later that Amoco had abandoned a refinery similar to Neodesha's in 1982 just outside Kansas City, Missouri, in a suburb called Sugar Creek. Studies showed there were 850,000 cubic yards of contaminated material at the Sugar Creek site. "That's enormous," Scheussler said. "And what's here?" After it inherited the Sugar Creek site, BP negotiated damage claims with affected property owners and convinced the state to allow "natural attenuation"—or leaving the pollution in place—saving the company at least $250 million in cleanup costs. "That to me was absolutely incredible," Scheussler said. "It looks to me like the same thing was happening down here. And it bothered me."

A star witness for the plaintiffs testified on a cool, rainy Monday in mid-October, when some of the jurors were beginning to show signs of wear from more than six weeks of testimony. Daniel B. Stephens, a hydrologist from Albuquerque, New Mexico, had been hired by the city to conduct an independent assessment of the pollution soon after the lawsuit was filed. Stephens, a slender man in his fifties wearing a blue suit, white shirt, and red-and-blue-striped tie, described how benzene, arsenic, and a host of other dangerous chemicals sat in a thick, gel-like layer over the groundwater and permeated the water itself beneath three-quarters of the city. Everyone in the courtroom, many of them Neodesha residents, leaned forward on the edges of their seats as Stephens used a laser pointer to pinpoint the worst "hot spots" on maps of the city showing the plume below.

Stephens then described how the clay soils above the groundwater were filled with cracks and fissures that allowed toxic gases from the plume to seep into the air above, including the indoor air of homes and businesses without much insulation. And he told how heavy rains and melting snows would push the groundwater table upward, forcing oily contaminants to leach into pools on the ground or even into the basements of leaky buildings.

As Stephens testified, the attorneys for BP appeared to be slumping in their seats. Their demeanor did not improve when Edgar showed a videotaped deposition of an engineer who had been hired by BP to assess the Neodesha site. The consultant, Brad Simmons, explained that he had been trained to write a contamination report with "a craft for persuasion," to help lead the reader to a preordained conclusion. Attorney Edgar asked Simmons what conclusions he had been asked to make for BP. Simmons stopped short of saying he had been told to lie, but the looks on some of the jurors' faces suggested they thought that is exactly what he had done—with some shaking their heads in disbelief—as Simmons described how reports could be drafted and tests could be limited to produce results that minimized the extent of the pollution.

Another BP consultant, Jenny Phillips, was afflicted by memory lapses when she was on the stand. Edgar showed her reports that she had written about the contamination, describing low levels of benzene and other contaminants in the water, or arguing that no protective barriers were needed at the site because there was no threat of pollution escaping into the air. Later studies proved the reports dead wrong, but Phillips accepted no responsibility, saying she could not remember writing them.

A state health official responsible for ensuring there was no risk to the public in Neodesha also withered under Edgar's questioning. Kurt Limesand was asked what steps the Kansas Department of

Health and Environment had taken to test, monitor, and reduce threats from the pollution. Limesand could not think of anything the state had done in the community, independent of BP. Basically, he said, the regulators had relied entirely on the company's assessments.

Another KDHE official, Pam Chaffee, the site manager for the state in Neodesha, admitted under questioning by Edgar on a Monday that she had met for eight hours the day before with Lamb, BP's lead attorney, to prepare for her appearance that day.

Edgar wrapped up Neodesha's case by showing a videotape from a bus tour BP had arranged for the advisory board it handpicked in 2003. Lucille Campbell, one of the board members, had fortuitously brought a friend with a video camera along on the tour, and the result was some of the strongest evidence suggesting that the company misled the community. On the tape, BP's Lloyd Dunlap is seen kneeling on the front seat of the bus, displaying a map of Neodesha and the old refinery. As the bus arrives at a fenced location, Dunlap describes how a lightning bolt struck a storage tank there in 1968, spewing all its chemicals onto the ground. That was the primary source of the pollution beneath the site, he says.

At the end of the tour, the camera turns toward Lucille Campbell, seated at the back of the bus. "If you went into a home and you saw a husband with a smoking gun in his hand and his wife is lying dead on the floor from a gunshot wound, it is pretty clear what happened," she says. "But if the death of this woman affects the economy of the town in which they live, and the husband asks you to help him bury his wife and not tell anyone, you would be an accessory. This is what has happened at this oil site."

Lucille Campbell did not get to see herself on the videotape that day in court. She was at Mercy Hospital in Independence, Kansas, at the bedside of her husband, who was suffering from lung cancer. Bob Campbell was a tough man, a former gandy dancer for the Frisco Railroad, who did the hard physical labor of building and maintaining the tracks. He later found employment as a state high-way worker, but an accident in the early 1970s forced him to retire from there with a disability, and he went on to work for the Kansas Bureau of Investigation, helping track down drug dealers. But Bob couldn't overcome a lifelong cigarette habit, and the chemotherapy prescribed for him in the summer of 2007 left him severely weak-ened in the autumn. In mid-October he was hospitalized with pneumonia and Lucille remained with him every single day, until he was finally sent home with oxygen supplies after Thanksgiving.

By that time BP was presenting its case, and Lucille managed to drive to Erie in her van equipped with a wheelchair lift—even through a nasty ice storm one day—to provide a show of support for the town in court. She heard testimony from Tammy Brendel, a BP environmental manager who had overseen Amoco's refinery shutdowns in Sugar Creek, Missouri, and Casper, Wyoming, both of which resulted in lengthy legal battles with communities con-cerned about the limited cleanups. And she watched in disbelief as Norm Bennett, a longtime Neodesha resident who worked for BP, argued that the company had done all it could to protect the com-munity from the contamination.

Lamb called fewer witnesses than the city, after having labored to discredit the plaintiffs' case through his cross-examinations of their testimony. But one surprising witness he brought forward was

DeWayne Prosser, the Baptist preacher who had kicked Lucille Campbell out of his congregation for advocating a lawsuit years before. Prosser described how Mrs. Campbell and her supporter, salvage yard operator Rick Johnson, had once brought a Geiger counter to a house in Neodesha that Prosser used for a youth ministry. "They began to circle the building, and the Geiger counter was going buzz, buzz, buzz, off the chart," Prosser said, apparently forgetting that Geiger counters make a clicking sound when radioactivity is detected. Prosser was concerned about Johnson's discovery, though, and asked the city, the county, and the state to conduct tests at the property, all of which turned up negative, he said. He also asked BP to have a contractor test soil and air samples, and he provided a separate air sample to an independent lab that did tests by mail, and all of those also found no contaminants present at the house.

Lamb then asked Prosser for an assessment of Lucille Campbell, even though she had not been called as a witness—partly because she lived outside the boundaries of the contaminated plume, so she was not eligible for the class action—and was known only to the jurors as someone who apparently pushed for the city to file the lawsuit. "She published a newsletter she called NEAT, and it was anything but neat as far as its structure was concerned because basically what she did was photocopy everything negative about BP that she could ever find—" Prosser testified. Edgar interrupted, saying that Prosser's answers were nothing more than speculation and opinion, but Lamb had made his point to the jury—this was a community that was out to "get" BP.

As the case wound down in mid-December, some courtroom observers sensed that the BP attorneys, especially Steven Lamb, seemed resigned to losing the case. The fire they had displayed in August had become as cold and icy as the Kansas winter. But after all, Lamb and his team would be paid handsomely regardless of the outcome, and any damages would come from BP's substantial coffers. All that remained, many felt, was for the jury to decide how large those damages should be.

And so the exhausted attorneys from both sides arrived early on the frigid morning of Tuesday, December 18—a day later than planned because of a fierce weekend snowstorm—to present their closing arguments. Despite the weather, the courtroom was packed, overflowing, in fact, into the hallways of the small courthouse. It seemed that half of Neodesha had braved the elements to make the twenty-five-mile drive to Erie, to witness the culmination of the town's epic battle with one of the world's largest corporations.

The final arguments were almost anticlimactic, after forty-six witnesses had provided sixty-nine days' worth of testimony.

Edgar's case was simply that BP had put profits over the safety of Neodesha, and lied to avoid a cleanup. "To BP, Neodesha means nothing—except as a place where they have to spend money," he said.

BP used the state health department "as a bodyguard" to protect it from a more costly cleanup, Edgar said, noting that the KDHE site manager, Pam Chaffee, had admitted meeting with BP's attorney for eight hours on the day before she testified in the case. "She wouldn't meet with me, the representative of the people of Neodesha she was supposed to be trying to protect," he said. "But she went and met with the polluter."

Edgar informed the jurors that while BP was fighting Neodesha's demands for a cleanup, the company was settling more than two dozen lawsuits in the Kansas City suburb of Sugar Creek claiming health damages from pollution surrounding the abandoned refinery that Amoco had earlier accepted as its responsibility in a federal consent decree. One case in Sugar Creek that went to trial resulted in a jury award of more than $13 million for a cancer victim there, he said.

"When I first undertook this case, I was driving into Neodesha and I saw a little girl about the age of my granddaughters running down Granby Street, her books from school in her hand, running to the mailbox in a cloud of dust. And I knew that there was a risk that that cloud of dust contained contaminants. It was a picture like a faded-out kind of picture with her and the cloud of dust. I have never forgotten that little girl, never forgotten that little girl. These are kids, not just adults. Those are kids that are being exposed to this. We have to protect these children," Edgar said.

"Couple of Sundays ago when I was at church the kids were getting their Bibles and I saw the kids going to the—up to the front like we have seen them walking down the streets of Neodesha, different outfits on, different hair, little kids," Edgar went on. "Then one kid, one child came out of the wings pushing another child in a wheelchair. We have to protect these people. You have to protect these people. We already have one child with AML [acute myeloid leukemia] in Neodesha. How many more is it going to take before we get the message to BP? You can give them that message.

"It is you who will determine whether the largest producer of oil in the United States, the eighth-largest corporation in the world, the second-largest oil company in the world, gets to continue to use Neodesha in southeast Kansas as its toxic waste dump," Edgar told the jury. "The people of Neodesha want their town back. They want to be free of the curse of contamination."

Lamb summed up BP's case by saying the city's lawsuit had more to do with "politicians and money"—a small town trying to capitalize on BP's oil wealth. "They want theirs," he said. "They want their share."

He seized on Edgar's references to children as gratuitous grandstanding. "We heard about a little girl running in the dust in Neodesha, and we transitioned to the kids because we need to think about the kids. And of course we love the kids, we love the children," Lamb said. "From there we went right to the little girl in the wheelchair. And you know what? I know something about the little girl in the wheelchair, too, because when my daughter Ashley was born from the age of two to nineteen she sat in a wheelchair. I pushed her to the doctor, I pushed her in the park. And when she died two years ago I held her hand as her light left the room and went to a better place. And I went to church and I delivered the eulogy.

"That is not what this case is about. There is no little girl running in the dirt in this courtroom. She didn't testify here, her mom didn't testify here. Her dad didn't testify here. There is discussion about there is a child we know with AML in Neodesha. Do you think that we wouldn't want to know that? . . . No. It's not in the evidence."

Lamb argued that KDHE officials had made it clear in their testimony that the agency was not a rubber stamp for BP. Risk assessments in Neodesha all met state standards, and in one case was accepted by a judge. The cleanup system installed by BP on the old refinery site was working, and would eventually remove all contaminants from the groundwater, only at a slower pace than the city wanted, Lamb said.

"And probably most important, Tammy explained that, you know what, we'll test anybody anywhere anytime for anything," he said, referring to testimony by KDHE's Tammy Brendel. "Soil, vapor, water, what do you want? We'll do it. How do we know that happened? Pastor Prosser came in and said, 'Yeah, I want it.' Okay, fine. We went and did it. City hall asked for it. We did it. Anybody that wants it, we'll do it."

Lamb concluded by summarizing testimony from the regulators, scientists, and company officials BP had called, who each outlined steps that were taken to monitor the site and protect public health. The city's case, he said, "is all built on this concept of fraud and the only way for it to work is if you believe that everybody, everybody, is lying. And that's just not the case."

With that, the twelve jurors gathered in a small room in the Erie courthouse and, finally, after four grueling months, began deliberating on Thursday, December 20. Following a two-week break for Christmas and New Year's, the jury returned to the courthouse on Monday, January 7, 2008.

After just three days of deliberations, the jury filed back into the courtroom that day and stunned the people of Neodesha: BP was not guilty on all counts. The dozens of residents who attended the announcement, including Lucille Campbell, sat in silence as the jurors filed out of the courtroom. Outside, Edgar and his team of attorneys were practically speechless, saying only they were deeply disappointed.

"We are shell-shocked," Rochelle Chronister, a former state legislator from Neodesha who attended much of the trial, told reporters after the verdict was announced. "It was just like we were at a funeral. We just don't understand what happened."

One juror, Ann Atchison, said the decision was fairly easy. "We didn't find anything that showed us any wrongdoing by BP," she said. "Nothing stood out as any wrongdoing on BP's part."

BP attorneys quickly left town, leaving it to company headquarters to issue the reaction. "The jury's verdict reflects the findings of an independent government study that confirmed that contamination from a former Amoco refinery that was closed thirty-seven years ago poses no apparent public health hazard," said a statement from BP spokeswoman Valerie Corr.

The lawyers for Neodesha returned to Kansas City to regroup and put together an appeal. In Neodesha, those who had been hopeful a cleanup would soon be under way lost faith that anything would ever be done about the contamination.

"There's few words to describe the shock felt in the courtroom at Erie on January 7 as the verdict was read," Lucille Campbell wrote a few days later for a Christian newspaper in southeast Kansas called *Good News*. She recounted the entire saga in a lengthy article, from her first awareness of the pollution to the trial in Erie that abruptly ended an eight-year struggle. Her husband Bob was back in the hospital after having trouble breathing again, and Lucille spent most of her days and nights at his side, sleeping when she could in a chair. It was hard not to be depressed. "Yet I'm standing up to BP's lie," she said.

The Kansas Department of Health and Environment, after a public whipping during the trial for relying heavily on BP to assess the pollution problems in Neodesha, made a show of support for the community by hosting meetings to hear residents' concerns. Lucille attended one of the sessions and became even more discouraged. "I hate to say it, but the KDHE people could have almost been mistaken for a group of BP people," she said afterward.

Adding insult to Neodesha's injury, news came from Sugar Creek outside Kansas City in March 2008 that BP had settled the twenty-eight remaining lawsuits claiming health damages from pollution from the refinery that Amoco had operated there from 1904 to 1982. Only one of the thirty-one cases filed against BP had gone to trial, and that resulted in a $13.3 million jury award, plus undisclosed punitive damages, for a man whose wife died from leukemia. Two other cases were settled earlier for undisclosed amounts: One for a man diagnosed with non-Hodgkin's lymphoma, and another for a man who had lived in Sugar Creek from his birth in 1981 until the age of five, was diagnosed with non-Hodgkin's lymphoma and leukemia in 1985, and underwent chemotherapy and radiation treatment as a child.

Then, eight and a half months after the Neodesha trial ended, the hopes that had been dashed in the community were suddenly revived. Judge Daniel Creitz, acting on a motion filed by the city, issued an eighty-six-page ruling on September 22 throwing out the jury's January 7 verdict and ordering a new trial strictly on the amount BP should pay to clean up the town.

The judge denied many of the city's motions for a new trial, including appeals arguing unprofessional conduct by BP's attorneys, admission of testimony by witnesses who primarily offered opinions, and possible misconduct by one of the jurors. However,

Creitz said the instructions to the jury were either misunderstood or improperly spelled out. He said the question of liability for the contamination was not really one for the jury to decide but that it was a matter of law that BP was responsible for the pollution.

"Given the defendants' admissions and evidence here, no rational juror could return a verdict stating that [BP was] not guilty of contaminating the groundwater underneath Neodesha," Creitz wrote in his ruling. "The contaminants in this case are some of the most dangerous known to mankind."

Creitz also said the contaminants had spread beyond the refinery site and that BP's efforts to remediate the groundwater had been insufficient to address the full extent of the pollution.

The city's lead attorney, John Edgar, felt exonerated. "I had confidence in our case, and believed that ultimately as a result of our motion and the court's ruling on it, that we would take a step closer to justice—and I believe that is what happened," he said.

Creitz gave BP ten days to appeal his ruling, and that is exactly what the company did. The appeal went first to the state appeals court, which in turn passed it on to the Kansas Supreme Court to make a final decision. Neodesha's battle with BP would continue for at least another year and possibly longer.

chapter seven

FADED GREEN

Tony Hayward's first full year as BP's chief executive began in January 2008 with news that another worker had been killed in an accident at the company's Texas City refinery, where payments for claims from the March 23, 2005, explosion were soaring past $2 billion.

William Joseph Gracia, a lifelong employee at the refinery, died after a 500-pound metal lid blew off a water-filtration tank and struck him on the head. Gracia, fifty-six, was also burned on his hands, face, and head in the blast, but the cause of death was ruled as blunt-force trauma. The incident occurred in the refinery's ultra-cracker unit during a restart of equipment, always a dangerous time.

Just a day before the fatal accident, Gracia had told his wife that he was thinking about retiring after thirty-two years at the plant. His daughter, Nicole Pina, described Gracia as a genuine patriarch who cooked the family's meals, kept everyone smiling, and was proud of his two children and five grandchildren. "He was just everything, and now we're trying to put it back together. It's impossible. Every day you wake up, and it's never going to be the same," she said.[1]

Pina said she promised her father on his deathbed that she would not let BP sweep his death under the rug.

Gracia was actually the third worker to die at the Texas City refinery since the 2005 explosion that killed 15 and injured 180 others. A contract worker was crushed to death in July 2006 when he became caught between a pipe and a mechanical lift, and another contract employee was electrocuted in June 2007. The death of Gracia in yet another explosion brought the Chemical Safety Board back to Texas for a new investigation and called into question Hayward's claims that he planned to completely revamp the safety culture at BP.

About a month after Gracia's death, an internal BP document surfaced showing that safety and environmental problems were actually mounting during Hayward's reign. In the six months after he took over for Lord John Browne in May 2007, leaks and spills at BP operations increased slightly from the previous six months, from thirty incidents to thirty-three. The problems were concentrated at BP's refineries in Texas, Washington, and Indiana, and on its pipelines in Alaska, the "Major Incidents Reporting" memo showed. Hayward had warned after he became CEO that it would take a long time to turn around the company's accident record, and now it appeared to be getting worse instead of better.[2]

The refinery problems were also dragging down BP's bottom

line. The company reported a 22 percent drop in profits in 2007. It estimated a net profit of $21.2 billion for the year, a healthy showing, but one well below rivals ExxonMobil, which broke corporate records with $40 billion in profits in 2007, and Royal Dutch Shell, which reported $27.6 billion. Hayward tried to put the best face on things by pointing to a fourth-quarter surge in revenues that pushed up dividends by 25 percent. But overall he had to admit that the company's 2007 performance was "very poor," especially in a market where global oil prices were steadily rising and had leaped past the unprecedented level of $100 a barrel.

Since replacing Browne, Hayward had made streamlining and simplifying his watchwords at BP. One of his first acts was to hire the consulting firm Bain & Co. to "hold up a mirror" to BP and tell him what he had inherited from Browne. "I was gobsmacked," he told *The Sunday Times* of London later about the consultant's report. "They said, 'You are the most complicated enterprise we have ever come across.'"[3] In response to that and to growing pressure from his board and shareholders for better performance, Hayward announced an austerity program for 2008 that would eliminate more than 5,000 jobs and reduce company overhead costs by at least 15 percent, while also increasing stock dividends. (The cuts did not apply to Hayward's income, though. He received a $2 million cash bonus for his work in 2007, on top of his $1.4 million salary.)

At the same time, the heat was rising on all of Big Oil from consumers angry about the relentless rise in gasoline prices, which were approaching $4 a gallon in America for the first time. Executives from BP and other oil companies became a regular punching bag for members of Congress demanding to know why drivers were being pummeled at the pump. The answer was simple, but unsatisfying—there was not enough oil to meet growing demands.

Hayward had started at BP as a twenty-five-year-old geologist on an oil rig, and he spent years in the 1990s leading exploration efforts in Colombia and Venezuela. He considered himself first and foremost an oil man, and he longed to get the company back on track to focus on its core business—petroleum.

One way of doing that was by expunging the trappings of Lord Browne and his varied outside interests. Hayward had all of the artsy paintings removed from BP headquarters and replaced with photos of drilling platforms and oil workers. He also had the office of BP Chairman Peter Sutherland moved down a floor at headquarters and placed adjacent to his, signaling his desire to work more closely with the company's board.

Hayward put his philosophy into action, making overtures in early 2008 to leaders in war-torn Iraq about helping increase production from the massive Rumaila oil field on the Kuwait border. He also pushed investments in the potentially lucrative but environmentally risky oil sands of Canada, a venture Lord Browne had considered ill-advised.

But for all his efforts to find ways to boost production, Hayward kept getting pulled back to problems at BP refineries, which combined had lost $192 million in 2007 and were still draining corporate coffers for upgrades and maintenance. The Texas City refinery, in particular, continued to haunt BP, with the seemingly endless litigation over the 2005 disaster constantly putting the company on the defensive in the U.S. and British press.

Debate over the U.S. government's proposed fine of $50 million for the explosion raged in a federal courtroom in Houston, as Judge Lee Rosenthal heard from a parade of victims about how their lives had been altered by the tragedy. David Senko, the J. E. Merit Constructors supervisor of the fifteen workers who died, told the judge that the penalty was minuscule for a company that could easily have afforded better protections for workers. "The $50 million being talked about, it's not even a decimal point on BP's scale," he said. "Any one of many, many people could have prevented it from happening." Eva Rowe, who lost both her parents in the blast, also made an appeal. "If the purpose of punishment is to give incentive to a wrongdoer to change their ways and do the right thing, this agreement utterly fails," she said. "Fifty million dollars is less than one month's worth of profit for this one BP plant."[4]

A retired federal judge, Paul Cassell of Utah, offered his services pro bono to the families of those who died in the refinery and appealed to Rosenthal not to accept a $50-million fine. "I think the victims have a very strong argument that this was not a good plea agreement," Cassell said.[5]

Plaintiffs' attorneys contended BP's fine for the fatal blast should be at least $400 million, and perhaps go as high as $3 billion. Rosenthal said she would consider all the arguments, including BP's plea that the fine was more than adequate considering the company was paying more than $2 billion to compensate victims. While her deliberations continued—delayed by an appeal of Rosenthal's ruling that the government had not violated victims' rights in negotiating the settlement—the House Energy and Commerce Committee opened an investigation into how the plea deal with BP was worked out, putting the Texas tragedy in the national media spotlight again.

Attorneys for victims planning another civil case against BP drew renewed attention to the accident, too, by questioning Lord Browne in a deposition about how Texas City was managed from BP headquarters. The deposition was recorded by telephone in April 2008, with Browne in London and plaintiffs' attorney Brent Coon in Houston. It proved to be more entertaining than informative, however. Browne set the tone by reciting his full name as "Edmund John Phillip Browne, the Lord Browne of Madingley," and advised Coon, "I am also entitled to this—the address 'the Right Honorable' and the suffix 'knight.'" Under questioning from Coon, Browne acknowledged meetings with President George W. Bush, Vice President Dick Cheney, and former British prime minister Tony Blair, and mentioned acquaintances with celebrities Gwyneth Paltrow, Hugh Grant, and Elton John, including a visit to the singer's Venice apartment. But he denied any knowledge of safety problems in Texas City, and said he saw nothing unusual there during two visits to the refinery before the March 2005 accident.

News of Browne's testimony did not sit well with Senator Edward Kennedy (D-Massachusetts), chairman of the Senate Health, Education, Labor and Pensions Committee and a longtime champion of workplace safety laws. Kennedy demanded that the Labor Department demonstrate it was addressing safety issues at all U.S. refineries, putting more pressure on BP and others to make sure they were complying with regulations. Several months later, the Occupational Safety and Health Administration hit BP with citations for fifteen serious safety violations at the plant, including problems that contributed to the death of William Gracia in January. OSHA had already fined BP $21 million for violations leading up to the March 2005 accident.

Coon pressed forward with another case filed by ten plaintiffs injured in the Texas City explosion and stunned potential jurors even before the trial started in May 2008 by announcing he would seek $950 million in punitive damages from BP. "We believe the only way a company like BP is going to get the message is to hit them in the pocketbook," he said.

By June, BP was scrambling to settle the remaining eighty-nine claims still pending from the accident. The judge overseeing the cases, State District Judge Susan Criss in Galveston, canceled all but one trial related to the explosion that was left on her docket. The last remaining case was dismissed for lack of evidence in September, ending the state litigation three and a half years after the accident.

The federal case against BP still languished, though, stalled by an appeal to the U.S. Supreme Court that victims' rights had been denied by prosecutors who negotiated the proposed settlement with BP. The appeal was rejected in July 2008, and Judge Rosenthal pledged to move toward a decision on the settlement, although with a word of caution in October. "I can't make that plant safe," she said. "All I can do is accept or reject this plea."

Rosenthal did the former in March 2009, agreeing to accept the plea deal and end prosecution for the Texas City explosion four years earlier. BP was ordered to close the case with a fine of $50 million. "It just doesn't make sense," a tearful Eva Rowe said after the judge's ruling was announced. "It seems so unfair that someone isn't punished and they just get to write a check."[6]

A Justice Department spokesman acknowledged that the plea agreement would not ease the pain of the accident's victims. But, said department spokesman Andrew Ames, "It demonstrates that the federal government takes seriously its mission to prosecute those who knowingly violate the nation's environmental laws."

By this time, Tony Hayward had far bigger headaches to deal with at the helm of BP. After having reported stunning record profits of nearly $10 billion for the second quarter of 2008, BP began seeing its share prices spiral downward as financial markets around the world began to collapse in the autumn. Oil prices were plummeting as well, to less than $70 a barrel, or half of where they had been a year earlier.

BP profits set another record in the third quarter of 2008, pushing over $11.5 billion, but the growth in the bottom line brought a new problem: Conservatives in Britain and the nation's biggest union slammed BP for maintaining high fuel prices to bolster profits when oil prices had dropped significantly. "While the poor and vulnerable cannot make ends meet or heat their homes, the greedy oil companies are banking money faster than they can count it," snapped Tony Woodley, joint general secretary of Unite, representing 1.5 million workers in Britain and Ireland.

BP also had its own foreign policy crisis unfolding with Russia, as its partners in a joint oil production company known as TNK-BP became openly hostile to the British managers and threatened to kick them out of the country. The Russian operations accounted for about a quarter of BP's oil output worldwide, so the threats, which BP viewed as an attempt to force it to pull out of the venture, carried huge financial implications for the company. When the

Russian government announced in early 2008 that it would take
control of a gas deposit in Siberia owned by TNK-BP, BP's market
value dropped 8.1 percent.

The chairman of the Alfa Group, the Russian side of TNK-BP,
accused BP Chairman Sutherland of using "Goebbels propaganda"
to bully its partners and attacked Hayward in a press conference
shown live on Russian television. "There is a good English word—
arrogance—we have sensed this kind of condescension for a long
time," said Alfa chairman Mikhail Fridman. Robert Dudley, the
CEO of BP's side of the operations, was repeatedly harassed by Russian authorities about taxes and other issues, and at one point was
forced to leave Moscow and manage the company from elsewhere.[7]

Hayward's response to the global economic problems was to pull
back harder on the reins. In early 2009 he cut investment in refining and marketing by 20 percent, delaying some expansion plans
and upgrades needed to make BP more competitive in those areas
with ExxonMobil and Royal Dutch Shell.

Cognizant of the growing economic crisis, Hayward ordered a
low-key recognition of BP's one hundredth birthday on April 14,
2009. At the company's annual meeting two days after the cente-
nary, Hayward said he viewed the looming recession as an opportunity to set the company apart from its competitors. "We have a
world-class resource base and our improving track record for exe-
cuting projects gives us real confidence for the future," he said.

The BP board of directors had another major worry at that year's
annual meeting. Chairman Peter Sutherland, a former Irish attorney general and member of the European Commission, had been
planning to leave the board that day after more than a decade as its

leader. The only problem was that BP hadn't been able to find a suitable replacement, forcing Sutherland to stay on until a successor was named.

The extension of Sutherland's chairmanship came at an awkward time for him and didn't please some shareholders. Sutherland had been forced to resign as a nonexecutive director of the Royal Bank of Scotland after the U.K.'s second-largest bank nearly collapsed in 2008, saved only by a government bailout. BP board member Tom McKillop had been chairman of the Royal Bank and was pressured to step down from his BP position earlier in April, and some shareholders said Sutherland should, too. "When Sir Tom's job is untenable, I do not understand why your job is not," an investor stood up and told Sutherland at the annual meeting.[8]

Sutherland would not have even faced the question if BP's earlier search for a chairman had been more fruitful. The company thought it had the perfect candidate in January—Paul Skinner, a longtime executive with Royal Dutch Shell who had become chairman of the British mining company Rio Tinto. But Skinner was forced to abandon the BP opportunity in February when Rio Tinto investors threatened a revolt over Skinner's plan to cover some of his company's massive debt with $19.5 billion in cash from a metals group controlled by the Chinese government.

BP headhunters turned next to John Bond, chairman of telecommunications giant Vodaphone, and two former chairmen of the U.K.'s Financial Services Authority, Adair Turner and Howard Davies. All three declined the BP chairmanship. By April, in what the British press described as one of the longest executive searches in corporate history, BP had gone through a candidate list that included leaders at Rolls-Royce, Thomson Reuters, and National Grid, all without finding anyone to take the job.[9]

It wasn't until June that the board settled on the head of a Swedish telecommunications company, Ericsson CEO Carl-Henric Svanberg, as its next chairman. Business analysts across the continent were left scratching their heads. Svanberg's chief claim to corporate fame was as an aggressive cost-cutter, though he was known to be a social and environmental activist. Among his roles, he was a member of the external advisory board for the Earth Institute at Columbia University, a position he shared with the rock singer Bono and the chairman of the United Nations Environment Programme's Intergovernmental Panel on Climate Change, R. K. Pachauri.

The appointment of Svanberg, who had little background in the oil industry, gave Hayward a freer hand to push the company toward the future he had long envisioned for it. Critics sneered that it was simply a strategy of less green (cleaner energy) and more black (as in black gold).

Hayward had set the stage for cutbacks in alternative energy programs early in 2008 when he indicated that BP's investments in solar, wind, and other renewable energy sources could eventually be put up for sale. "We intend to grow this business predominantly for its equity value," Hayward told a group of energy investors in February 2008.[10] He estimated the value of the company's alternative energy portfolio to be between $5 billion and $7 billion, adding, "As we go forward, we will be looking at how best we can realize that growing value for our shareholders."

The head of BP's alternative energy division, Vivienne Cox, tried to dispel rumors that swelled after Hayward's remarks that her operation was going on the market. "That doesn't mean we're going to divest it, in case anyone gets the wrong idea," she told reporters in late February.[11]

BP did continue investing in green projects even as it seemed to be contemplating a future without them. In the spring of 2008, the company announced it would spend $8 billion over the next decade on alternative energy, including $1.5 billion in the year to come, or 7 percent of its capital spending. Plans were announced for a $1 billion project in Canada to find cleaner ways to extract natural gas, for extensive research on biofuels in the United States, and for installation of scores of solar units to provide electricity to U.S. utilities.

Hayward and other BP executives also maintained the activist stance on climate change that had been the company's official position since Lord Browne defined it in 1997. BP joined with 139 other multinational corporations in late 2008 urging the United Nations to develop strong standards for reducing greenhouse gases, and Hayward endorsed strict limits that would be met with emission trading programs in all nations. "The ultimate objective is a global cap-and-trade system, but that's probably a little way off," Hayward told an energy conference in February 2009. "The best place to start is at the national level."[12]

BP's chief scientist, Steven Koonin, also was a vocal advocate for cap-and-trade systems to put a price on sending carbon emissions into the atmosphere. But Koonin, who would later be tapped for a top post in the U.S. Department of Energy in the Obama administration, lamented in September 2008 that the developed world was slow to give up its dependence on carbon-based fossil fuels. "We're trying, [but] it's not easy to change things," he said in a September 2008 talk at the Massachusetts Institute of Technology. "You can't cut off the present. . . . Deployment of energy innovations is very hard because of entrenched interests."[13]

And BP was doing everything it could to respond to those interests. In a move that set Hayward apart from Browne more than any other, the company announced in December 2007 that it was acquiring half of a Husky Energy operation developing Canadian oil sands in the western province of Alberta. In return, Husky, based in Calgary, Alberta, would take a 50 percent share of BP's refinery in Toledo, Ohio, which would be expanded to process the heavy Canadian crude. Two months later, BP announced a similar deal with Husky that included plans for a $3.8 billion expansion of the Whiting, Indiana, refinery in anticipation of more oil flowing down from the north.

The oil sands had been proven to contain at least 170 billion barrels of crude—and very likely held much more—and were fast becoming the source of 20 percent of the U.S. domestic oil supply. BP was the last of the oil majors to get into the action, having been kept out by Lord Browne's opposition to the expensive and energy-intensive process of extracting petroleum from the deep layers of largely untouched earth.

Environmental groups howled at the announcement, angry that the oil company that had promised to move "beyond petroleum" was now becoming engaged in the least environmentally friendly way to produce it. Sparing no hyperbole, U.K.-based Greenpeace called it "the biggest environmental crime in history." Added Ann Alexander, senior attorney for the Natural Resources Defense

Council: "Tar sands crude oil is dirty from start to finish. It's bad enough that [BP is] fouling our natural resources here in the Midwest, but it's completely destroying them up in Canada. There are good sources of energy we can turn to that don't involve turning entire forests into a moonscape."[14]

Most of the oil-rich bitumen in Alberta lies deep underground, and must either be mined or pumped out using hot steam in a so-called in-situ process. Mining the sands leaves the largest environmental footprint: Acres of fields and forests are replaced with deep open pits, and the extraction of bitumen leaves huge piles of toxic tailings and millions of barrels of poisonous sludge that is dumped into vast, man-made lakes. There are concerns about extensive water pollution, cancer in downstream communities, mountains of toxic residues, and emissions of greenhouse gases in quantities three times greater than conventional oil-drilling operations because of all the heavy equipment that is used.

Hayward insisted that BP would not engage in mining the oil sands, but would rely on the in-situ process. "BP has never been in the strip-mining of the tar sands and never will be," he would say several years after the operations began. "We are focused on so-called steam-assisted gravity drainage, which is much more akin to conventional reservoir engineering . . . therefore the environmental footprint on the ground is no more or worse than normal oil or gas operation."[15]

In-situ operations involve pumping hot steam deep into the ground to melt the bitumen so it can be extracted. But the process is extremely energy-intensive; it takes about 250 cubic feet of natural gas to produce one barrel of bitumen from oil sands. The Canadian Indigenous Tar Sands Campaign, a network of tribal nations opposed to oil sands development, estimated that the process results in carbon-dioxide emissions three to five times greater than conventional oil production.

North American environmental groups and indigenous peoples were not the only ones raising concerns about BP's plans in Canada. Socially responsible investment groups, which had committed to BP shares during John Browne's tenure as CEO, began to reconsider whether BP still met their criteria as an environmentally responsible corporation. "Oil sands development offers some of the worst life-cycle environmental impacts of any fossil fuel—emitting nearly triple the greenhouse gas emissions of traditional oil extraction," said Miles Litvinoff of the Ecumenical Council for Social Responsibility in April 2008. "Prior to BP's announcement in December, we had understood that our company would not pursue tar sands development due to the heavy carbon footprint of both the operations and the end product. We fear the implication that BP is retreating from an excellent strategic position designed to exploit the long-term shift away from high-carbon fuel sources and question whether this may undermine BP's future competitiveness."[16]

There were other signs that BP was putting the Beyond Petroleum campaign permanently on ice. As the global recession deepened, the company's alternative energy programs were the first to be targeted for deep cuts. In November 2008, BP eliminated virtually all plans for wind-power systems and other renewable projects in the U.K., saying it would shift the bulk of its $8 billion alternative energy program to the United States, where greater tax incentives were available to make projects more attractive. BP also pulled out of Britain's competition to design the first power plant that could capture carbon emissions and store them underground.

Then in June 2009, Vivienne Cox, who had insisted the previous year that BP had no plans to divest the alternative energy division that she headed, resigned to spend more time with her family and BP divested the division. The budget would be cut from $1.5 billion to between $500 million and $1 billion, depending on how much ongoing projects could be scaled back in the United States. BP also closed a Maryland plant that had been manufacturing solar-energy equipment, eliminating 320 jobs, and said it would focus on lower-cost solar projects in China and India.

The moves even startled Lord Browne, who was now overseeing a renewable energy fund in London for the New York–based equity firm Riverstone. "I don't think that BP has rowed back from tackling climate change and I very much hope that it does not move away from my position," he told the *Sunday Telegraph* in February 2010. "I have always believed the oil majors—like all companies—should be doing more. But companies need to have the right carrots and sticks in place from governments."[17]

BP was clearly focusing more on the central issue for Hayward and BP shareholders—oil. The steep decline in prices on the world market added to the pressure for development of new reserves to help maintain the financial commitments of jittery investors. The drop in crude prices to around $60 a barrel also put a squeeze on many drilling contracts that had been negotiated when prices were above $140 a barrel. Payments being made to contractors like Transocean, which leased massive drilling rigs in the Gulf of Mexico, were based on wells being found that would recover oil at 2007 prices, so speed in bringing the wells online was critical to cutting costs.

BP invested about $30 billion in developing new oil and gas resources in the United States from 2005 to 2009, with most of that investment concentrated in the deep waters of the Gulf of Mexico. Drilling was progressing in nearly a dozen deepwater leases held by BP, including one of its most promising prospects ever at the so-called Tiber field about 250 miles southeast of Houston. The discovery of a deposit containing at least 3 billion barrels of oil was announced in September 2009 after the *Deepwater Horizon*, a rig leased from Transocean, had drilled the deepest well in history, to more than 35,000 feet below the ocean surface. "What today's announcement proves is that BP is a very, very successful explorer," said Irene Himona, an oil and gas analyst at Exane BNP Paribas, when the Tiber find was announced. "They've opened up the whole area for discoveries."[18]

BP's *Thunder Horse* platform, the world's largest drilling rig that had been delayed by problems found after a hurricane nearly sunk it in 2005, also was finally producing the equivalent of 300,000 barrels of oil and gas per day at its site 150 miles southeast of New Orleans. Another BP rig, the *Atlantis*, was producing about 200,000 barrels of oil daily from leases held about 170 miles south of New Orleans beneath 7,000 feet of water.

The *Atlantis* had gone into operation in 2007 after it, too, was delayed by the fierce hurricane season in 2005, and there were some concerns that BP may have rushed it into development. In 2009, a whistle-blower who had worked for a BP contractor provided a congressional committee with e-mails dated in August 2008 showing that BP did not have complete drawings and records of the rig's design and operations on board the *Atlantis,* a potential serious violation of federal regulations. Members of the House Natural Resources Committee demanded that the Minerals Management Service look into the allegations in February 2010, and less than two months later a BP ombudsman who investigated the whistleblower's information said it had been substantiated.

There were other signs that safety was a growing concern on BP drilling rigs in the Gulf. An internal report written in December 2007 by Richard Morrison, BP's vice president for Gulf of Mexico production, said there had been ten events with "high potential" for adverse consequences in BP's Gulf operations that year. "As we enter the last two weeks of 2007, we are experiencing an unprecedented frequency of serious incidents in our operations," Morrison said in an e-mail with his report. "We are extremely fortunate that one or more of our co-workers has not been seriously injured or killed."[19]

Nonetheless, BP aggressively pressed forward with deepwater exploration. In March 2010, it bought 240 more leases from Devon Energy Corporation in the area where the Tiber field had been discovered in 2009. Excitement about future prospects in the Gulf grew even stronger at the end of March, when President Barack Obama announced he was considering a plan to open more areas in the Gulf and along other U.S. shorelines to offshore drilling.

And in April 2010, BP executives were preparing to announce yet another major discovery, in an area about fifty miles southeast of Venice, Louisiana, which it dubbed the Macondo Prospect.

chapter eight

"NIGHTMARE WELL"

THE CALM WATERS AND DARK silence of the sea fifty miles from shore in the Gulf of Mexico belied the tension inside the control room of the *Deepwater Horizon* drilling rig in the evening hours of April 20, 2010.

The rig was sitting a mile above leased ocean bottomlands known as Mississippi Canyon Block 252, also called the Macondo field after a fictional South American town in the 1967 novel *One Hundred Years of Solitude* by Colombian writer Gabriel García Marquez. The book tells the story of Macondo's rise and fall over six generations, before it is wiped off the face of the earth by a devastating hurricane. The name might have been an omen for the holder of the lease, BP, and its minority partner, Anadarko Petroleum Corporation.

BP was leasing the rig from Transocean, the largest offshore drilling contractor in the world and one that had made history just seven months earlier by striking oil for BP more than 35,000 feet below the Gulf's surface, one of the deepest wells ever built. That well, in a deep-sea field about 250 miles southeast of Houston known as the Tiber Prospect, also was drilled by the *Deepwater Horizon*, a massive mobile platform roughly twice the size of a football field.

When it announced the Tiber find in September 2009, BP called it "a giant oil discovery," with some speculating it could contain as much as 3 billion barrels of oil. Now, after reaching down through 5,000 feet of water and drilling another 13,000 feet into the Macondo field, BP officials believed they were on the verge of making another exciting announcement. All the signs pointed to oil and gas deposits that would produce millions of barrels of crude, and possibly tens of millions, after Transocean's crew finished the well and capped it until a permanent platform could be put in place for production later.

The first part of the well had been drilled by another Transocean rig, the *Marianas*, which had to return to port for repairs after sustaining damage from Hurricane Ida in November 2009. The *Deepwater Horizon*, the star of Transocean's fleet of more than 130 mobile platforms, was moved into position to complete Macondo starting on February 6, 2010.

BP had a contract to lease the *Deepwater Horizon* at a daily rate of nearly $500,000, and the company had budgeted more than $500,000 additional for crew, supplies, equipment, and support operations, putting the cost of drilling the well at more than $1 million a day. The initial goal was to complete the Macondo well in seventy-eight days for about $96 million, but BP had made clear to Transocean it hoped the job could be done in fifty-one days. With the

Marianas having worked for a month drilling about 4,000 feet into the bottom of the Gulf, the expectation was that the *Deepwater Horizon* would go the remaining 9,000 feet and complete the well in twenty-one days, or by the beginning of March.

It soon became apparent in February that BP's schedule was ambitious. As the drilling proceeded more than two miles below sea level, the crew discovered on numerous occasions that it was losing some of the heavy mud that was pumped into the drill casing to counter pressure from below when oil, gas, or fluids were encountered. It was clear that weak formations deep in the rocks and salt were allowing mud to leak out, and the casing had to be reinforced nine times, slowing the drilling progress.[1]

Pressure increased in the well hole in early March and there were a series of "kicks" indicating that things were not under control and that gas could surge up at any time, endangering the rig. The drill became stuck at that point and had to be cut and replaced with a new drill that would veer off in a different direction. "We are in the midst of a well control situation on MC [Mississippi Canyon] 252 #001 and have stuck pipe," BP manager Scherie Douglas wrote in a March 10 e-mail to federal regulators in New Orleans. "We are bringing out equipment to begin operations to sever the drillpipe, plugback the well and bypass." The disruption would delay progress on the well by about two weeks.

By mid-April—six weeks behind BP's schedule—the drill reached a major oil deposit and excitement began to grow on the rig. There were still indications of unstable conditions and growing pressure at the bottom of the well, and BP had to seek permission from regulators to make some last-minute design changes. But things were looking up, with the end of the well preparation in sight. Still, there was as much anxiety as there was a sense of relief among the crew on the *Deepwater Horizon*. "This has been [a] nightmare well which has everyone all over the place," BP engineer Brian Morel said in an April 14 e-mail to a colleague.

On April 20, Transocean's managers on the rig and supervisors for a handful of contractors gathered shortly before 11:00 A.M. to discuss the final plans for sealing the well so the *Deepwater Horizon* could move on to its next drilling job—one that had been scheduled to start forty-three days earlier. BP was paying a high price for keeping the rig longer than expected, with cost overruns on the Macondo well now more than $40 million past the original budget of $96 million.

Transocean's crew and contractors wanted to move carefully, given all the problems they had encountered to this point. According to sworn testimony later by witnesses to the meeting, no sooner had the group begun going over the final process when BP manager Robert Kaluza took charge and laid out an accelerated timetable for cementing the well. An argument ensued between Kaluza and Transocean's senior manager on the rig, Jimmy Harrell, but BP would get its way. "This is how it's going to be," Kaluza said, and Harrell "reluctantly agreed."[2]

Kaluza's orders were to begin removing drilling mud from the well and replacing it with seawater later that day, allowing the cementing contractor, Halliburton, to install a final plug below the ocean surface that would close the well until it was ready for production. The plan would short-circuit a number of tests meant to ensure that no gas or oil was leaking into the well, but the crew went along, feeling it was BP's call on how it wanted to complete construction of its well. "I guess that is what we have those pinchers for," Harrell was heard to grumble after the morning meeting, making reference to the steel clamps on a device over the well called a blowout preventer, that was supposed to squeeze the pipe shut if gas and oil began spewing upward.

Not long after the meeting, a group of BP and Transocean executives arrived by helicopter for a tour of the *Deepwater Horizon* and a reception with its captain, Curt Kuchta, to celebrate seven years without a serious accident on the rig. The company officials were in high spirits, but the mood among workers in the drilling room below deck was tense. Some felt that BP was rushing to finish the well, without taking adequate precautions against a blowout.

The worries increased around 5:00 P.M., when the crew did a "negative pressure test" on the well after some of the drilling mud had been removed, to see if there were signs of leaks into the well. The first test showed there could be a leak, with mud being pushed up an escape valve and sputtering to the surface. When the test was repeated there were more signs of pressure coming into the well. As the workers huddled in the drilling-floor "shack," or control room, the tour of BP and Transocean executives arrived for a quick look, guided by Harrell. As the tour left the room, the crew asked Harrell to check out the pressure test results. Harrell said they seemed fine, suggested that the escape valve on top of the blowout preventer be tightened a bit, and left the room to return to his guests.[3]

At least one supervisor of the drilling operation for Transocean, Wyman Wheeler, wasn't convinced everything was right, and asked his replacement for the next shift starting at 6:00 P.M., Jason

Anderson, to do an assessment of his own. Anderson looked over the test results and concluded the pressure readings were not that unusual.

BP's own manager on the day shift, Kaluza, also was perplexed by the 5:00 P.M. tests. He took the results to Donald Vidrine, a more experienced manager who was scheduled to take over on the night shift. Vidrine suggested that another test might help resolve matters.

A little before 8:00 P.M., the crew did another test to determine if there was pressure coming in from outside the well casing and if there was sufficient pressure inside the well to hold the cement seal in place when the rest of the drilling mud was removed. The results just muddied the waters even more. Devices on a small tube running up from the well showed no pressure, while gauges inside the main pipe showed above-normal pressure. Since the tube and the pipe were linked, there should have been no difference in the readings. Vidrine apparently concluded that the smaller tube must have been clogged since it showed no pressure at all, and told Kaluza to call BP supervisors in Houston and tell them everything was a go to remove the rest of the mud and place the final seal in the well.[4]

Within a couple of hours, some of the 126 workers on the rig began seeing and hearing and smelling signs that something was terribly wrong. Those who had finished their twelve-hour shifts for the day were resting comfortably in their rooms, relaxing in the theater, working out in the exercise room, or enjoying some of the other amenities in what they called their offshore "hotel." Many in the crew were at the end of their twenty-one-day tours and looking forward to the next three weeks at their homes on land.

Transocean's chief mechanic, Douglas Brown, was in the engine control room around 9:45 when he heard an "extremely loud air-leak sound" just as gas alarms on the rig went off. Two of the engines near him began to rev up.[5]

Mike Williams, in a workshop nearby, heard the same thing. "I hear the engines revving. The lights are glowing. I'm hearing the alarms," he told the CBS program *60 Minutes* a few weeks later. "I mean, they're at a constant state now. It's just 'beep, beep, beep, beep, beep.' It doesn't stop. But even that's starting to get drowned out by the sound of the engine increasing in speed. And my lights get so incredibly bright that they physically explode. I'm pushing my way back from the desk when my computer monitor exploded."[6]

At 9:50 P.M., Transocean's second-ranking manager, Randy Ezell, was watching satellite television in his room when he received a distress call from Stephen Curtis, working in the drilling room. "We have a situation," Curtis said to Ezell. "Randy, we need your help." Curtis said methane gas was gushing up the well pipe and filling up the room. The shift supervisor, Jason Anderson, was trying to close a cover over the drill casing to block some of the gas and was attempting to activate the blowout preventer, Ezell was told.[7]

Vidrine, the BP senior manager, was in his office doing paperwork when he also got a call from the drilling room. Mud was shooting up the well, which was out of control, the worker reported. Vidrine started rushing toward the deck and noticed that mud and seawater were raining down from the sky. Just then, a massive explosion occurred.[8]

At that moment, Curtis and Anderson were almost certainly killed instantly as the drilling room erupted in a ball of fire, workers said later. Eight others in the room—six more workers for Transocean and two with a mud-engineering firm, M-I SWACO—were also apparently incinerated. Dale Burkeen, a crane operator, was trying to scramble down a catwalk to safety when a second explosion occurred and he was thrown more than fifty feet to his death on the deck, workers said. His body was never recovered, though, amid the pandemonium that ensued.

Wyman Wheeler, the Transocean supervisor who had been worried that something was wrong with the well when the pressure tests were taken earlier, was in his room packing for a return flight home the next day when the explosions began ripping apart the rig. As Wheeler tried to scramble toward the deck, the door of his room blew in on him, breaking his shoulder and leg. Other workers who were rushing past pulled him out of his room and carried him to the deck.[9]

Mike Williams, who had heard the engines revving from his workshop, also was blown backward as he reached for the door to flee and was pinned against the wall. After he crawled out, another door blew off its hinges and he angrily pushed it aside and moved toward the deck. By the time he arrived, many crew members were already scrambling into lifeboats or, worse, jumping ten stories down to the ocean below. More explosions shook the entire rig. Asked later by correspondent Scott Pelley on 60 Minutes to describe the blasts, Williams said, "It's just take-your-breath-away type explosions, shake-your-body-to-the-core explosions. Take your vision away from the percussion of the explosions."[10]

The power of the initial blast knocked Micah Sandell from the seat of his crane, and he scrambled down a spiral staircase toward the main deck. About ten feet from the bottom, the second explosion lifted him over the railing and he fell to the deck, but he was able to get up and start running. "Around me all over the deck, I couldn't see nothing but fire," he later told *The Wall Street Journal*. "There was no smoke, only flames."[11]

On the bridge of the rig, which was not yet damaged by the explosion or flames, Captain Kuchta and about a dozen crew members and managers from BP and Transocean hurriedly tried to assess the situation. But the rig's navigation operator, Andrea Fleytas—one of only three women on the floating platform—wasn't waiting for instructions. She grabbed the radio and shouted, "Mayday, Mayday! This is *Deepwater Horizon*. We have an uncontrollable fire." When the captain heard the distress call, he snapped at Fleytas that she did not have authority to issue it. Fleytas apologized.[12]

The top BP and Transocean managers, Vidrine and Harrell, joined the captain on the bridge and noticed that the control panel showed the drilling crew had closed the cover over part of the drill in an effort to block or divert the gas, but it obviously had little effect. Chris Pleasant, the operator of the blowout preventer, raced to activate the device, but before he pushed the button he could see there was no pressure showing in the hydraulic system, which meant the pinchers to squeeze the well shut would not work. Pleasant told the captain he was going to try to start the shears that were supposed to cut through the pipe and block the flow of gas and oil from the well, and though Kuchta told him to calm down and wait, Pleasant hit the emergency switch anyway. The indicator light changed from green to red, giving Pleasant a momentary hope that the shears had started, but then he realized the meter showed that no hydraulic fluid was flowing down to the blowout preventer. The last fail-safe measure on the rig had failed, and Pleasant knew it was time to leave.[13]

Kuchta's chief mate, David Young, had run to the deck to assess the fire, and by this time had raced back to report that only one crew member had suited up in firefighting gear, and the situation seemed hopeless. Kuchta agreed. "We had no fire pumps," he would tell an investigative board later. "There was nothing to do but abandon ship."

Still, Kuchta did not issue the order immediately, even as many in the crew were already scrambling into lifeboats or jumping over the side. Fleytas, the navigator, decided that someone needed to act. She flipped on the public address system and announced, "We are abandoning the rig."

Vidrine, the BP executive, made his way to one of the two enclosed lifeboats that were available to lower dozens of people off the rig. Inside, he heard crew members say the engine room had been gutted.

Douglas Brown, the Transocean mechanic who had been working in that room when the engines started revving up before the first explosion, had crawled out of the rubble and made it to the deck, and watched in horror as the rig's 242-foot derrick was engulfed in flames. All around him was total panic, with people running toward the sides, screaming and crying, he said later. Brown made it to a lifeboat, and a longtime friend who was counting heads was in such a state of shock that he couldn't remember Brown's name.[14]

Others, like crew member Matt Hughes, jumped over the side. Hughes, twenty-six, had been lifting weights when he heard the explosions, grabbed a life jacket and raced to the deck, only to find a crowd standing near a lifeboat still hanging by the side of the rig, uncertain of what to do. Hughes feared he would be burned to death if he waited and leaped over the railing to take an eighty-foot plunge into the water.[15]

Darin Rupinski, a navigation operator, told *The Wall Street Journal* that the scene was one of utter chaos. "One guy was actually hanging off the railing . . . people were saying that we needed to get out of there." Rupinski was going to help lower a lifeboat when a Transocean executive told him to wait until injured workers could be put on board. But while he waited, at least fifty people on the lifeboat started screaming at once for him to lower them down to the water. Some panicked workers just gave up and jumped.[16]

Not far from the burning rig was a support ship, the *Damon B. Bankston*, which had just delivered supplies to the *Deepwater Horizon* and off-loaded used drilling mud and other materials. The captain, Alwin Landry, heard the explosions and was watching the drama unfold when he noticed the reflective life vests of some crew members dropping into the sea. Landry turned his ship around and dispatched a small rescue boat, and within a short time his crew was plucking people out of the water and pulling people from the lifeboats.

There were fewer than a dozen people left on the rig at that point, including Captain Kuchta and Andrea Fleytas, the navigation operator who had sounded the command to abandon ship. The two lifeboats had already been lowered into the water, so the last survivors attached a large rubber raft to a winch, inflated it, and started climbing aboard. The injured Transocean supervisor, Wyman Wheeler, was helped onto the raft and Fleytas climbed aboard just as the descent began. The last stragglers, including Kuchta and Mike Williams, jumped over the railing.

Fleytas later recalled in an interview with *The Wall Street Journal* that when the raft hit the water, it wouldn't move. Crew members who were hanging on to it said they thought the heat from the rig was pulling the raft in, and a terrified Fleytas decided to jump off. "All I saw was smoke and fire," she said. "I swam away from the rig for my life." The rescue boat from the *Bankston* soon spotted her and pulled her aboard, and a crew member from the boat noticed that the raft was still hooked to a line tied to the rig. Someone passed a knife to one of the survivors, who cut the line, and the raft was towed to safety.

Within an hour after the first explosion, 115 survivors from the *Deepwater Horizon* were safely aboard the *Bankston*. Eleven crew members were missing, and though a search was conducted immediately, their bodies would never be found.

THE HUGE, UNCONTROLLABLE FIRE THAT raged on the *Deepwater Horizon* for more than a day after the accident put a tragic and very public exclamation point on more than a century of offshore drilling in the United States, a history that most Americans only think about when a disaster strikes.

The very first oil well dug off the U.S. coast—in 1896 in the Pacific Ocean about four hundred yards from shore near Santa Barbara, California—was abandoned after less than ten years of modest production and left behind a blackened beach, rotting piers, and unsightly derricks that remained until a tidal wave washed them away in 1942.

Exploration companies moved farther out to sea during the 1900s, but stayed in shallower waters close to shore because of the limits of drilling technology. The dangers weren't widely recognized until January 1969, when a Union Oil Company well blew out in the Santa Barbara Channel, causing an eight-hundred-square-mile oil slick and spoiling about thirty miles of beaches in Southern California. The spill led to tighter regulations and a ban on drilling along most of the U.S. coastline, pushing offshore oil and gas development almost exclusively to the Gulf of Mexico, where it was a mainstay of the economies in Louisiana and Texas.

Over the next four decades there were many smaller leaks, but no other major blowouts in U.S. waters that reached the size of the Santa Barbara spill. The *Exxon Valdez* tanker accident in 1989, which dumped 11 million gallons of crude on the pristine Alaska shoreline, kept the oil industry on the defensive for the rest of the twentieth century, however.

Oil production in the shallow waters of the Gulf remained fairly steady in the 1970s and 1980s, though a petroleum glut led to a collapse of world prices in 1986 that put a hold on most expansion plans. By 1990, barely over 1 million barrels of oil per day were being produced from offshore wells. But at the same time a new trend of exploring in deeper waters was taking hold that would set the stage for growth in the offshore industry by the end of the century.

The first successful well in water more than a thousand feet deep was drilled by Shell Oil Company in 1975, in the so-called Cognac field in the Mississippi Canyon about a hundred miles southeast of

New Orleans. The wellhead, 1,025 feet below sea level, barely set the deepwater mark, but it was one that stood as a record for Shell, the leader in deep-sea technology at that time, until it drilled a well in water 1,330 feet deep at the same field in 1989.

The industry moved into deeper and deeper waters in the 1990s, accelerated by passage of the Deep Water Royalty Relief Act in 1995, which allowed the government to waive royalties on wells drilled below two hundred meters of water, or depths greater than 656 feet, for the next five years. The law worked as intended: By 2000, production from wells between a thousand and five thousand feet surpassed production from wells in shallower waters, and oil output in the Gulf of Mexico reached a peak of more than 2 million barrels a day in 2002. Companies set their sights on the potential for vast deposits of oil and gas at much greater depths, and rigs were developed that could extend drills to the ocean floor in waters more than a mile deep.

The expansion of ultradeepwater drilling started raising concerns among some regulators. The federal agency charged with overseeing oil and gas development on public lands, the Interior Department's Minerals Management Service (MMS), began sounding alarms regularly in its reviews of exploration plans in the Gulf.

In one report issued in 2000 assessing a plan by Shell Deepwater Development Company for drilling in nearly three thousand feet of water, the MMS said one of the greatest risks would be a fire on the rig caused by a surge of gas upward through the pipes. The result could be a blowout thousands of feet below the surface that would be very difficult to control, the agency said. The likelihood of extensive damage to the environment, including fish kills and destruction of wetlands, would be very high from such a spill, the report said.

"In the event that a subsea blowout occurs, the intervention that would most likely be employed to regain control of the well would be the drilling of a relief well," the MMS predicted. "Drilling an intervention well could take anywhere from thirty to ninety days." But the agency also concluded the chances of that happening were low, based on the industry's past performance in the Gulf. "Potential impacts from an accidental release of oil from a high volume blowout is of concern; however, the historical database indicates that it is rare for such a pollution event to occur," the report said. "For the period from 1971 to 1995, based on 24,237 wells drilled, there were 17 well blowouts that resulted in the release of oil (0.07 percent probability)."

Concerns about deepwater drilling began to fade within the government when President George W. Bush, a Texas oil man, took office in January 2001 with an expressed mandate to expand domestic energy sources. Four months after moving into the White House, Bush signed an executive order requiring the MMS to expedite the process for issuing oil and gas drilling permits, and the race was on among oil companies to tap into the deep waters of the Gulf of Mexico.

Bush's vice president, Dick Cheney, formed a task force to develop a new national energy plan, with the oil, gas, coal, nuclear, and electric industries all given seats at the table. The Republican-controlled Congress put many of the group's recommendations into law with passage of the Energy Policy Act in 2005 that provided a host of new incentives for energy development, including more royalty relief for exploration and production in the Gulf. Even so, by the middle of Bush's second term, with supplies squeezed by Middle East tensions and growing oil demand in developing nations, the cost of crude skyrocketed on the world market and gasoline prices in the United States soared past $4 per gallon. Politicians responding to angry constituents pushed even harder for development of domestic supplies. "Drill, baby, drill" became the battle cry, led by Republicans such as Alaska Governor Sarah Palin, who landed a spot as running mate to Senator John McCain (R-Arizona), in the 2008 presidential election.

The political pressures hit directly on the Minerals Management Service, an agency with conflicting mandates. It was charged with regulating drilling to safeguard energy workers and the environment, but it also collected the revenues from oil and gas leases on public lands, taking in an amount for the federal government second only to the taxes paid into the Internal Revenue Service. To keep the royalties—and the oil and gas—flowing, the MMS tended to give the industry plenty of leeway in its exploration and production operations.

The cozy relationship between the industry and its regulators erupted into a full-blown scandal in 2008, when the Interior Department's inspector general reported that MMS employees overseeing oil and gas drilling in the West were literally sleeping with the energy officials they were regulating. An investigation also found regular drug parties between regulators and industry lobbyists and a free flow of gifts to government officials, including lavish meals, trips, and tickets to sporting events. "This all shows the oil industry holds shocking sway over the administration and even key federal employees," said Senator Bill Nelson (D-Florida), when the investigation's findings were revealed.

Later reports by the inspector general found that oil companies operating in the Gulf were sometimes allowed to fill out their own inspection reports in pencil, and MMS employees in the field would ink over them with a pen before sending them to agency managers.[17] One of the reasons was simply an inability to keep up with the workload from fast-growing drilling operations, as the MMS struggled with a tight budget and lack of expertise in an industry that was becoming more complicated the deeper it went into the uncharted bottomlands of the Gulf.

"In the deep water—that's precisely when these agencies need more resources to deploy the most extensive analysis that is available to us," Lynn Scarlett, a former deputy secretary of the Interior, told the *Los Angeles Times* a few weeks after the *Deepwater Horizon* accident. "Unfortunately, they don't have those resources available to them. That may ring hollow to the American public. But I have sympathy for agencies that are asked to do so much."[18]

The week after President Barack Obama took office in January 2009, his new Interior secretary Ken Salazar pledged that the administration would completely overhaul regulation of the oil and gas industry with protection of the public interests taking precedence over the needs of the companies. "There's a new sheriff in town," he said.

Salazar, a former U.S. senator from Colorado, went directly to the MMS office in Lakewood, Colorado, that had been at the center of the scandal during the Bush administration and laid down the law. "Our agenda for reform will reach every part of this department," he told the staff there. "But it will also send a loud and clear signal to the special interests outside of this department who have become accustomed to the 'anything goes' attitude in Washington over the last eight years. The 'anything goes' will end. And this department, and the Minerals Management Service, will lead the way in ending it."

Later that year, the MMS proposed to strengthen its oversight of offshore energy development by replacing a voluntary safety program that had been in place since 1994 with a requirement that companies have independent audits done on their equipment at least once every three years, accompanied by more frequent on-site visits by MMS inspectors. The agency said its review of offshore incident reports from 2001 to 2007 showed there had been 41 deaths and 302 injuries in 1,443 accidents over seven years of drilling in the Outer Continental Shelf (OCS). "The MMS believes that if OCS oil and gas operations are better planned and organized, then the likelihood of injury to workers and the risk of environmental pollution will be further reduced," the agency said in its proposal to abandon voluntary audits and make them mandatory.

Industry officials were virtually unanimous in their opposition to the proposal. "While BP is supportive of companies having a system in place to reduce risk, accidents, injuries, and spills, we are not supportive of the extensive, prescriptive regulations as proposed in this rule," BP's vice president for Gulf of Mexico production, Richard Morrison, wrote to the MMS in response to the proposed rules. "We believe industry's current safety and environmental statistics demonstrate that the voluntary programs implemented since the adoption of API RP 75 have been and continue to be very successful." API RP 75 is the "Recommended Practice for Development of a Safety and Environmental Management Program for Offshore Operations and Facilities" written by the American Petroleum Institute, the oil industry's chief lobbying arm in Washington. RP 75 was just one of nearly a hundred recommended practices or standards for offshore drilling that had been drafted by the API and written into federal regulations with few, if any, changes.[19]

After all the howls of protest from the industry, the MMS dropped the regulatory proposal in late 2009 and the voluntary safety program for offshore drilling remained in place.

A few months later, at the end of March 2010, President Obama proposed an expansion of offshore drilling, saying the administration would consider opening new areas of the Gulf of Mexico and OCS lands off the coasts of states that agreed with lifting a federal drilling ban that had been in place for decades. One of Obama's aims was to offer a concession to Republicans who were blocking Democratic bills in the Senate to address climate change, in hopes of getting enough votes to enact controversial legislation that had already passed in the House. But Obama and his aides also had clearly bought into the oil industry's arguments that offshore drilling had proven safe enough to be expanded without fear of environmental consequences. In a comment that would come back to haunt him just a few weeks later, Obama offered a defense of his proposal to allow more drilling on April 2, 2010, saying, "Oil rigs today generally don't cause spills. They are technologically very advanced. Even during Katrina, the spills didn't come from the oil rigs, they came from the refineries onshore."

Had the government been monitoring drilling operations more

closely in the Gulf, it might have been alarmed by what it found, just on the *Deepwater Horizon*.

An audit of the rig's safety and environmental programs had been done in 2007 by a Norwegian risk-management company, Det Norske Veritas (DNV), and found that maintenance was overdue by months on some critical safety equipment. But DNV issued a new five-year certification, as it often did in voluntary audits performed under contract with the rig's owner. The Coast Guard, which oversees safety on commercial vessels operating in U.S. waters, could have done an inspection on the rig, but already had its hands full keeping watch on U.S.-flagged ships. Transocean, headquartered in Switzerland, had registered the *Deepwater Horizon* in the Marshall Islands, which allows ships operating under its flag to hire private firms for safety inspections. In most cases for foreign-flagged vessels, the Coast Guard accepts the results of those audits without conducting inspections on its own. And much to the surprise of the Coast Guard and the MMS when it was revealed later, the private audit on the *Deepwater Horizon* did not include a thorough examination of the all-important blowout preventer, the fail-safe piece of equipment meant to stop a wellhead leak before it gets out of control.

Transocean's recent record in the Gulf might have raised eyebrows among U.S. regulators as well. After it acquired the Houston drilling company GlobalSantaFe in 2007, Transocean was responsible for 73 percent of all incidents reported in the next three years, according to a study by *The Wall Street Journal*, even though it owned less than half the rigs operating on the Gulf. Nevertheless, the MMS awarded Transocean with its top safety award in 2008, and BP, which was responsible for about 12 percent of Transocean's business, was a nominee for the SAFE award in early 2010. (After the April 20 blowout, the MMS decided not to issue the award.)[20]

The MMS, responsible for approving drilling plans and inspecting operations on all federal leases, also appeared to give wide berth to BP at the Macondo well field.

An exploration plan for the project filed with the MMS in February 2009 was insistent that any type of spill from the operation was "unlikely" or "virtually impossible," and even if one did occur, BP promised it "has the capability to respond to the appropriate worst-case spill scenario." The agency approved the plan less than two months later, with one word of warning based on its review of data on the well field. "Exercise caution while drilling due to indications of shallow gas and possible water flow," MMS officials told BP when the plan was approved.[21]

Around the same time, in April 2009, the MMS issued a "categorical exclusion" for BP's Macondo plan, meaning it would not have to meet the usual federal requirements for an extensive review of the project's possible environmental impacts. The exemption from the National Environmental Policy Act (NEPA), a landmark 1969 law requiring environmental impact statements for any project using federal funds or lands, was not that unusual. After President Bush ordered an expedited process for issuing oil and gas permits, the MMS began granting NEPA exemptions for drilling operations at the rate of nearly one per day.

MMS officials argued that under the streamlined process, they were required to act on permit requests within thirty days, leaving little time to review the sometimes lengthy exploration proposals filed by oil companies. A federal court ruling in 2008 said the

MMS did have authority to take more time on permit reviews if it decided an exploration plan was incomplete, but the reality was that the agency had been virtually rubber-stamping plans—and granting between 250 and 400 "categorical exclusions" for them annually—since the early years of the Bush administration. President Obama's Interior secretary, Salazar, ended the practice in 2009—after BP had already received its exclusion for the Macondo well.

The MMS also appeared to be giving only cursory reviews to emergency response plans submitted by oil companies in advance of drilling projects. A 582-page response plan filed by BP for drilling in the Mississippi Canyon included long passages that were identical to plans prepared for other projects, including some proposed by other companies. One section of BP's response plan raised the possibility that "seals, sea otters, and walruses" could be affected by an oil spill in the Gulf, although none of those animals resides in southern climes. The company clearly had cut-and-pasted material from response plans for Arctic drilling projects, and the flaws were never noticed by the MMS when it approved the response plan in July 2009.

In the last week before the blowout occurred, the MMS received a flurry of requests from BP for permits that would allow changes in the well design previously approved of by the agency. While last-minute alterations in a complex drilling operation were not ususual, one aspect of BP's requests was almost unprecedented—it asked for three different revisions within twenty-four hours, a sign that it was scrambling to address problems with the well. All the permit changes

BP requested were approved by the MMS, in some cases within minutes.

On April 14, BP notified the agency that it needed to use a smaller pipe than planned earlier at the bottom of the well, because its nearly ten-inch diameter pipe would not fit into the hole that tapped into the oil reservoir 18,360 feet below sea level. The company now proposed using a single string of seven-inch pipe running all the way from the surface to the oil reservoir, rather than a double pipe with one encased in another—a method considered safer because it can prevent gases from shooting upward along the outside of the well. BP sent the single-pipe permit request to the MMS at 8:34 P.M. on April 14, and received approval the next morning at 8:13 A.M.

Less than two hours after receiving that permit on the morning of April 15, BP submitted another request, amending the plan again to have a tapered pipe going down the well, rather than a seven-inch pipe for the entire length. The MMS approved that permit in seven minutes.

Another change was requested in the afternoon, this time to correct an "inadvertent" error in a previous application that failed to mention a section of pipe already in the well. MMS approved that permit in less than five minutes.

A *Wall Street Journal* review of federal permit applications later revealed that BP's three requests for revisions within twenty-four hours were unique among all but one other of more than 2,200 wells built in the Gulf since 2004. The only other time that many changes occurred in a permit in that short a time was in 2005, for a well being drilled in less than fifty feet of water, the *Journal* reported.[22]

Whether any of the changes contributed to the explosion on the *Deepwater Horizon* had not been clearly determined by the end of 2010. Investigations by special commissions, federal agencies, congressional committees, and BP itself shed some light on possible causes of the accident, though, and there were many of them.

There was no doubt from the moment of the horrific explosion about what caused it. A massive gas bubble had surged up the more than three miles of pipe, spread over a portion of the drilling rig, and ignited, probably when it was sucked into one of the engines that crew members heard revving up just before the first blast.

How the gas got past the well seals and all the way up the pipe was the big question. Among the possible reasons were a cheaper, less secure well design; cracks in the well's cement liner; a weak seal at the bottom of the well; or some combination of all those factors. It was also possible the crew on the rig misread pressure tests that should have warned them that more work needed to be done to seal the well, before all the drilling mud was removed. Most troubling of all was the question of why the fail-safe device, the blowout preventer at the top of the well on the ocean floor, did not close the pipe and block the flow of oil and gas.

First there was the issue of the type of well BP built, which was a single pipe, or "long string" well, that many experts consider less reliable than a dual pipe, or a "liner tieback" well.

Evidence surfaced later that BP officials had vigorously debated this question, with some warning that the risk of using a long string was that gas could push up the outside of the pipe through the well hole if there were any gaps in the seals around it. "This would certainly be a worst-case scenario," BP drilling engineer Mark Hafle wrote in an internal report in 2009, obtained by *The New York Times*. "However, I have seen it happen so know it can occur."[23]

Hafle later denied in testimony about the accident that the use of a single pipe was a concern. "Nobody believed there was going to be a safety issue," he told a commission investigating the blowout. "All the risks had been addressed, all the concerns had been addressed, and we had a model that suggested if executed properly we would have a successful job."

BP's CEO, Tony Hayward, also told a congressional panel investigating the accident that a single long string of steel pipe is considered better for "the long-term integrity of the well," meaning it should be less prone to leaks when the well was producing oil in future years. "It was approved by the MMS," Hayward added.[24]

But in fact, other oil companies operating in the Gulf of Mexico did not use the long-string design for deepwater wells as much as BP, according to an examination of federal records by *The Wall Street Journal*. The study found that BP used the "long string" design on 35 percent of the deepwater wells it had drilled in the previous seven years, while Royal Dutch Shell used the design on only 8 percent of its deep wells and Chevron used it on just 15 percent of its wells.

The preferred method—even at BP on the majority of its wells—was the "liner tie-back" approach that involved cementing a pipe at the bottom of the well, then installing a smaller pipe inside the liner pipe that would draw oil from the well. The liner pipe also was connected to a tube that "tied back" to the top of the well, providing an extra barrier for any gas that flowed up the outside of the production pipe. Most experts consider the double-pipe method to be the "gold standard" for drilling deepwater wells, reflecting the fact that it is more expensive but provides greater protection against gas leaks.

Congressional investigators determined later that the safer, two-pipe system would have added between $7 million and $10 million to BP's costs and delayed the project another week. A BP engineer, Brian Morel, appeared to acknowledge that fact in an e-mail he sent on March 30, three weeks before the accident, saying that "not running the tie-back saves a good deal of time/money."

That decision infuriated leaders of the House Energy and Commerce Committee when it was revealed weeks after the accident. Committee Chairman Henry Waxman (D-California), and Representative Bart Stupak (D-Michigan), who chaired the panel's investigations subcommittee, told Hayward in a June 14 letter that BP had been told in mid-April that using a single pipe meant the only barrier to gas moving up the well was a cement seal below the wellhead. "Despite this and other warnings, BP chose the more risky casing option, apparently because the liner option would have cost $7 to $10 million more and taken longer," the lawmakers wrote. "The decision appears to have been made to save time and reduce costs."

Even BP's minority partner on the Macondo well, Anadarko Petroleum Corporation, questioned the decision to use a single pipe. Anadarko officials told *The Wall Street Journal* that although they used the long-string design on 42 percent of their deepwater wells in the Gulf, they went with the two-pipe system when drilling in unfamiliar geological conditions or in high-pressure gas and oil deposits, as was the case in the Macondo field. "It's not that long strings are unsafe, but they have to be under the right conditions," Darrell

Hollek, Anadarko's vice president of Gulf operations, told the *Journal*.

The long-string pipe meant there was no margin for error in the well's cement seals. Yet there were indications later that both BP and its cement contractor, Halliburton, may have taken steps in preparing the well that added to the risks of a gas leak.

A critical issue was the number of centering devices used to keep the pipe precisely located in the middle of the well hole; a pipe that angled even slightly could cause a gap in the cement seal that would allow high-pressure gas to surge upward through the well.

Halliburton's engineers had told BP that for a well that size—stretching more than 13,000 feet below the sea floor—it was important that at least twenty-one centering devices were placed at regular intervals on the entire length of the pipe. But in the weeks before the well was completed, BP decided to go with only six centering devices, according to documents obtained by the House Energy and Commerce Committee.

A Halliburton engineer in Houston, Jesse Gagliano, was alarmed by BP's plan. He warned BP engineer Brian Morel on April 15 that Halliburton's computer models showed that at least twenty-one centering devices were critical to getting a good cement job in the well. Morel responded in an e-mail later that day: "It's a vertical hole so hopefully the pipe stays centralized due to gravity."

But Gagliano was persistent and took his concerns to other BP supervisors. He finally convinced Gregg Walz, a new manager on BP's engineering team, that the Halliburton computer models were correct. "We need to honor the modeling," he told John Guide, BP's well team leader in Houston, in an e-mail. Guide was out sick, so BP executive David Sims intervened and ordered fifteen more centering devices and supplementary equipment to be sent to the *Deepwater Horizon* immediately.

Guide looked at the order the next day, April 16, and realized the devices that went to the rig that morning were the wrong size. "It will take ten hours to install them," he said in an e-mail to the drilling team on the rig. "We are adding forty-five pieces that can come off as a last-minute addition. I do not like this."

Brian Morel and Brett Cocales, two of the engineers on Guide's team, agreed with his position, even while expressing some reservations. "Even if the hole is perfectly straight, a straight piece of pipe even in tension will not seek the perfect center of the hole unless it has something to centralize it," Cocales wrote in an e-mail to Morel on April 16. "But, who cares, it's done, end of story, will probably be fine and we'll get a good cement job."

On April 18, Halliburton did another computer run of how a cement job might come out with only six centering devices on the Macondo pipe. The model showed there would be "severe risk of gas flow," a conclusion that Gagliano immediately sent to the BP engineers. However, the warning was included on page eighteen of a report that Gagliano attached to an e-mail, and none of the BP officials noticed it until after the April 20 explosion.[25]

Halliburton may also have been at least partially at fault if the cement seal did indeed break down and allow gas to burst through. Weeks before the spill, Halliburton conducted at least three different tests of the cement it had prepared for the Macondo well and the results all showed signs of instability in the mixture. Halliburton shared the results of only one of those tests with BP, on March 8, but there were no indications officials from either company ever discussed the problem, a preliminary report by the presidential commission set up to investigate the accident found later.

It was only on April 19, one day before the accident, that Halliburton received some data showing the cement should be able to hold under the intense pressure expected in the well, but the results were inconclusive and were not given to BP officials for their review. An independent lab hired by the presidential commission later tried to replicate tests on cement that Halliburton said was the same type used at Macondo; those tests showed the mixture to be unstable. "Halliburton (and perhaps BP) should have considered redesigning the foam slurry before pumping it at the Macondo well," the commission's staff concluded after analyzing the lab results.

If there were flaws in Halliburton's cement job, they might have been discovered before the final attempt was made to seal the well if BP had authorized a full battery of tests on the cement around the well casing. Tim Probert, a top executive at Halliburton, told congressional investigators later that two tests were conducted on the well that produced conflicting results. A third test—one considered the "gold standard" for determining whether a well is completely sealed—was canceled by BP less than twelve hours before the explosion, records later showed.

BP had hired a team from Schlumberger Limited, an oil services company that specialized in testing cement jobs, and had it waiting on the *Deepwater Horizon* on the morning of April 20 in case it was decided that a "cement bond log" would be needed. The test involves sending acoustic devices down the entire length of the well to produce visual images of any places where the cement had not fully bonded to the well casing, leaving cracks or fissures that could provide a pathway for gas. The test is highly reliable, but it requires nine to twelve hours to complete in a well as deep as Macondo.

Shortly after 11:00 A.M., BP told the Schlumberger employees their services would not be required and sent them home on a helicopter. The team was told that no cement had escaped from the well when it was poured, indicating that all of it had bonded and there was no need to test for defects. That was an assumption on their part, and although it was never stated, some have speculated that BP managers did not really want to know the results of a thorough test. If the bond log had revealed any flaws, BP might have had to order a new cement job that would take weeks to complete, and could cost at least $30 million, according to Tom McFarland, a well-cementing consultant from New Orleans who testified later about the possible causes of the accident.

McFarland, who examined detailed schematics of the Macondo well that had been provided by Halliburton's Tim Probert, was convinced there were flaws in the cement job simply because of the way the well was designed. There were tiny spaces in places where a larger pipe was fit over a smaller pipe as the well telescoped downward, and no cement job would ever be able to fill those cracks, McFarland told the Coast Guard and MMS commission investigating the accident. "It looks pretty on paper, but you can't accomplish that successfully and have a good cement job," he said. "The chance of getting a good cement job on that is nil."[26]

There was one other final test that might have shown if there were leaks in the well that also was cut off by BP. Drilling crews usually circulate mud from top to bottom and then back up through a well once it has been completed, and test the mud when it returns to see if it has absorbed any methane. For a well the length of Macondo, at 18,360 feet, it would have taken anywhere from six to twelve hours to completely circulate mud all the way up and down; BP ended the test on April 19 after only thirty minutes.

Perhaps some BP officials believed they could hedge on the final testing of the well because they knew there was a four-story device sitting above the wellhead on the ocean floor that would immediately cut off the flow of oil and gas if things started to get out of control. But if they had known the real state of the blowout preventer—and there were signs of potential problems long before April 20—leaders of the company's drilling team might have proceeded more cautiously.

BP's own records showed that at least three times in the weeks before the accident the blowout preventer was leaking fluids, most likely the critical hydraulic fluids that were needed to push the pinchers closed or the shears through the pipe to shut it off. A BP supervisor on the rig, Ronald Sepulvado, told the federal commission later that he reported a hydraulic leak to his managers and assumed the problem was being addressed. Mike Williams, the Transocean technician who was one of the last to jump off the burning rig after the explosion, also recalled in interviews and testimony later that there was a serious incident with the blowout preventer in March.

During a test of the device, when a key rubber gasket called an annular was closed to seal the valve at the top of the machine, a worker in the control room accidently bumped a joystick that pushed about fifteen feet of drill pipe down through the valve at high pressure. Drilling fluids that rose to the top of the well shortly after the botched test contained chunks of rubber, apparently from the annular. Williams showed them to a supervisor in the drilling shack who brushed off the matter, saying, "It's no big deal."

There must have been some concerns about the machine, though, because around the same time, as BP was struggling to control "kicks" in the well, company officials asked the Minerals Management Service to allow them to delay a pressure test on the blowout preventer that was scheduled to occur every two weeks.

Agency officials at first balked at the request, but under pressure from BP engineers who described "major concerns" with the well, the MMS allowed the test to be postponed. Mysteriously, when the blowout preventer was tested later, a much lower pressure than usual—6,500 pounds per square inch—was used on the device, BP records showed. The test was normally conducted at 10,000 pounds per square inch. The same lower pressure was used on all subsequent tests before April 20. Experienced drilling workers testified later they could not recall ever seeing such a dramatic change in testing procedures on a well.

Congressional investigators concluded there were a number of serious problems with the blowout preventer, including a dead battery, hydraulic system leaks, and the distinct possibility that the guillotinelike ram shears were not strong enough to cut through and close off the well, especially if they happened to hit one of the thick steel connecting joints that covered about 10 percent of the pipe.

An astonished Representative Bart Stupak noted that a Transocean report prepared when the blowout preventer was purchased in 2001 indicated there were 260 "failure modes," or potential problems that would require the device to be taken out of service and repaired. "How can a device that has 260 failure modes be considered fail-safe?" he asked at a hearing in mid-May.

A few months later a BP executive, Harry Thierens, added one more flaw to the list of problems in the blowout preventer. A review of detailed notes made when the device was last modified showed that workers may have installed a test pipe, rather than a pipe meant for actual operations, Thierens told the Coast Guard and MMS commission investigating the spill in August. "It would mean that the pipe rams could not be closed," he said. "I was frankly astonished that this could have happened."

When it completed its own internal investigation in September, BP laid some of the blame for the explosion on mistakes by its employees on the rig and on workers for Transocean and its contractors. The company cited as an example the final pressure tests that were conducted on the well just hours before the blowout, saying workers wrongly concluded that discrepancies in the results were not a problem. "The Transocean rig crew and BP well site leaders reached the incorrect view that the test was successful and that well integrity had been established," the BP report stated.

The company's conclusion was interpreted by some as an attempt to deflect responsibility for the disaster away from top corporate managers. It was considered especially galling because some of the very workers it accused of making mistakes—such as Transocean supervisor Jason Anderson, who had signed off on the pressure tests when he started his shift in the drilling room four hours before the explosion—were disintegrated in the blast. "BP is pointing fingers at those guys who were on the rig and they're not here to defend themselves," said Anderson's father, Billy Anderson, when BP released its report.[27]

Many government leaders and oil industry executives vented their anger publicly at BP in the months after the accident, often accusing the company of being reckless and putting profits over safety. Interior Secretary Salazar told a congressional hearing in June that "there were actions taken before April 20 which might have ended up creating this disaster, and perhaps no level of enforcement or regulation could ever have prevented that because of the recklessness that occurred here." BP's partner on the Macondo well, Anadarko Petroleum, issued a statement from its chairman, James Hackett, charging that "mounting evidence clearly demonstrates that this tragedy was preventable and the direct result of BP's reckless decisions and actions." Even President Obama used the word "recklessness" to describe BP's behavior in an address to the nation in June from the Oval Office.

But none of the officials directly attacked BP employees or their colleagues in other companies working on the Macondo well. No one believed any manager or line worker was so heartless and irresponsible that they would take unnecessary risks to save a small amount of money for a company that was making billions of dollars in profits. By and large, it was clear that employees of BP and its contractors were conscientious and professional, doing their life's work as best they could to provide petroleum to a world that was hugely dependent upon it.

"To date, we have not seen a single instance where a human being made a conscious decision to favor dollars over safety," said Fred

Bartlit, general counsel for the National Commission on the BP Deepwater Horizon Oil Spill and Offshore Drilling, more than six months after the accident. A veteran lawyer who had previously worked with oil companies on drilling issues, Bartlit said he did not believe any employee in the industry would do anything to risk the lives of their colleagues. "I've been on a lot of rigs and I don't believe people sit there and say, 'This is really dangerous, but the guys in London will make more money.' We don't see a concrete situation where people made a trade-off of safety for dollars.

"Anytime you are talking about a million and a half dollars a day, money enters in," Bartlit added. "All I am saying is human beings did not sit there and sell safety down the river for dollars on the rig that night."

What was troubling to investigators and overseers of the oil industry was that in virtually every instance where BP had a decision to make about the Macondo well, it chose the option that presented the lowest costs or saved the most time. There appeared to be a pattern deeply ingrained in BP's corporate culture where the costs and benefits of each action were weighed in isolation, without considering the bigger picture. The cascade of problems that led to the blowout in the Gulf seemed tragically similar to the series of actions and events that caused the refinery explosion in Texas and the pipeline problems in Alaska, both of which were blamed on a flawed "safety culture" at BP.

"What is fully evident, from BP's pipeline spill in Alaska and the Texas City refinery disaster, to the *Deepwater Horizon* well failure, is that BP has a long and sordid history of cutting costs and pushing the limits in search of higher profits," said Representative

Edward Markey (D-Massachusetts), a senior member of the House
Energy and Commerce Committee that investigated all three BP
disasters.

The leaders of the committee, Representatives Waxman and Stupak, reached much the same conclusion after just two months of
probing into the Gulf spill. "Time after time, it appears that BP
made decisions that increased the risk of a blowout to save the company time or expense," Waxman and Stupak wrote in a June letter
describing their findings about the accident to that point.

The cochairman of the federal agencies' commission that investigated the spill, Coast Guard Captain Hung Nguyen, said that there
appeared to be "many holes" in BP's approach to safety in the Gulf.
"What happened here?" Nguyen asked Harry Thierens, BP's vice
president for drilling and completions, at one commission hearing
in August. "Is it because the whole governance, in terms of the safety
net, there are so many holes in there, that these things fell through?"

Both cochairmen of the presidentially appointed commission
probing the accident raised similar concerns as their panel moved
toward writing a report that was presented to the White House early
in 2011. Bob Graham, a former U.S. senator from Florida, concluded in November that BP appeared to have a self-imposed deadline to complete its well by April 19 or 20, and pressure to meet it
may have led to some poor decision making. "There were a series of
almost incredible failures in the days and hours leading up to the
disaster," Graham said. His cochairman, former Environmental Protection Agency administrator William Reilly, echoed that concern.
"We are aware of what appeared to be a rush to completion," Reilly
said, concluding that there was "emphatically not a culture of safety

on that rig." After months of hearings on the accident, Reilly added: "I referred to a culture of complacency, and speaking for myself, all these companies we heard from displayed it . . . BP, Halliburton, and Transocean are in need of top-to-bottom reform."

THE MASSIVE FIRE ON THE *Deepwater Horizon* burned relentlessly for a day and a half after the first explosion occurred just before 10:00 P.M. on April 20, 2010. Firefighting boats continuously poured water on the rig with little effect, as the flames fueled by a steady stream of methane continued to shoot hundreds of feet into the air. As robotic submarines a mile below were maneuvered to try to force the shears closed on the failed blowout preventer and cut off the flow of gas and oil, the rig was rocked by more explosions early on the morning of April 22. A crisis manager bluntly assessed the situation in an e-mail: "Two explosions around 3:30-4:00 this morning & rig listing at about 35 degrees. High risk of sinking."[28]

At 10:22 A.M. April 22, the *Deepwater Horizon* did keel over and disappear into the Gulf, leaving stunned emergency crews watching in awe as a plume of steam rose and dissipated into the sky. The massive rig plunged to the bottom of the ocean and settled near the open wellhead, where oil and gas gushed through a mangle of twisted pipes.

chapter nine

THE BEAST

As soon as he heard the news about the *Deepwater Horizon*, Billy Nungesser, president of Plaquemines Parish on the southern Louisiana coast where many oil workers have their homes, realized it was an unprecedented calamity.

Nungesser could not recall a similar event in all his fifty-one years on the Louisiana delta. "We've had hurricanes and fires on the rigs, but I can't remember that we ever had this type of explosion and definitely not on this type of rig," he told *The New York Times* on April 21, 2010, the day after the accident. "This is one of the largest, deepwater, offshore drilling rigs."[1]

It would take a while for most Americans to grasp the full scope of the disaster, but those who were familiar with offshore drilling and had experience with oil spills were instantly alarmed by the images of the burning rig flashed around the globe. The commandant of the U.S. Coast Guard, Admiral Thad Allen, knew from four

decades of maritime experience that no one had ever dealt with a
well blowout a mile beneath the surface of the sea. And it was clear
from the massive ball of flames that there was gas and oil gushing
up from the deep at a furious rate.

Allen's colleagues at Coast Guard command centers in Washington and the Gulf provided a stunning assessment within twenty-
four hours of the explosion: "Potential environmental threat is
700,000 gallons of diesel on board the *Deepwater Horizon* and estimated potential of 8,000 barrels per day of crude oil, if the well
were to completely blowout," said a Coast Guard log obtained later
by the Center for Public Integrity. A daily leak of 8,000 barrels, or
336,000 gallons, would produce a spill the size of the *Exxon Valdez*
spill every month that it was not contained.[2]

The scene of the disaster, nearly fifty miles from the coastline of
Plaquemines Parish, also provided proof that this was no ordinary
accident. Within a few days a greasy sheen about a mile wide and
five miles long had formed on the Gulf waters. Scientists viewing
the satellite images had a sinking, helpless feeling when they realized the challenge of plugging a huge hole under more than five
thousand feet of water. "Everything about it is unprecedented,"
Christopher Reddy, an oil spill expert at the prestigious Woods
Hole Oceanographic Institution in Massachusetts, told the *Los Angeles Times*. "All our knowledge is based on a one-shot event. . . .
With this, we don't know when it's going to stop."[3]

Despite the dire warnings from the Coast Guard and outside
experts, the federal agency tasked with assessing marine data, the
National Oceanic and Atmospheric Administration (NOAA), offered an initial estimate that only about a thousand barrels of oil, or

42,000 gallons, were leaking from the well every day, but it would soon become clear that this was a very conservative figure.

On April 23, the day after the *Deepwater Horizon* sank, Coast Guard officials said privately they now feared the well could leak at a rate of more than 64,000 barrels per day—the equivalent of an *Exxon Valdez* accident every five days. Even the cautious scientists at NOAA took notice, though they still tried to keep their worst fears to themselves. "The following is not public," said a NOAA emergency response report on April 28. "Two additional release points were found today in the tangled riser. If the riser pipe deteriorates further, the flow could become unchecked resulting in a release volume an order of magnitude higher than previously thought."

The level of concern increased with each report from BP that it was unable to stop the leak. On the first day after the explosion, crews on response vessels at the site twice attempted to use robotic submarines to activate the blowout preventer sitting atop the well, both times without success. Although BP engineers weren't giving up yet, the situation appeared ominous.

The biggest problem with the blowout preventer was that hydraulic fluids needed to close the "shear ram" on the device were leaking out. The robots were used to try to seal the leaks, but every time one was closed another would spring up somewhere else. It took a week to make enough repairs to muster five thousand pounds of pressure on the blades of the shear ram, which was supposed to cut through the well pipe like a guillotine and close it off, but it was all for naught. "No indication of movement," BP said in an April 27 report on the attempt to close the blowout preventer.

When a scientist from the government's Los Alamos National

Laboratory, Scott Watson, arrived on the scene later to take X-ray images of the blowout preventer, BP's engineers were shown evidence that their earlier efforts to close the shear ram were only half successful. One of the two blades had cut through the pipe and been locked in place, but the other had not budged. "I don't think anybody who saw the pictures thought it was ambiguous," Watson said. With only one blade able to move, the blowout preventer was useless, and the well would continue to spew oil and gas through it unabated.[4]

Even before they had given up hope on the blowout preventer, the BP engineers knew they needed to quickly develop backup strategies for closing the well. The first idea was to build a gigantic steel dome to lower over the well in hopes of trapping much of the oil and funneling it to the surface. Workers began building three of the "subsea oil collection" systems at a plant in Port Fourchon, Louisiana, a few days after the accident, but it would take time to complete the forty-foot-tall boxes and get them to the site for placement on the ocean floor.

Meanwhile, the Marine Spill Response Corporation, set up by the oil industry after the *Exxon Valdez* accident in 1989, was mounting the biggest response effort in its history, according to its director, Steve Benz. The nonprofit company based in Herndon, Virginia, outside Washington had built up a response flotilla worth more than $100 million from fees paid by its member companies. The only problem was that the corporation's four hundred employees and fifteen cleanup vessels were scattered around the country, including at stations far from the Gulf in Alaska and Hawaii.

Overall, BP managed to muster thirty-two response vessels, including ships that could skim oil from the surface, within a week of the spill. The U.S. Navy added seven skimming systems and helped set up inflatable containment booms to try to block the flow of oil to the shore. More than forty miles of boom was laid in the week after the spill began, and nearly a hundred miles more was being assembled for deployment soon.

By the end of April, the response effort had grown to seventy-six vessels, eleven helicopters, six other aircraft, and a dozen remote-controlled machines and mobile platforms. Janet Napolitano, head of the Homeland Security Department that sent teams to the Gulf with employees from other federal agencies, estimated on April 29 that more than 1,100 people were working on the cleanup and had skimmed some 685,000 gallons of oil-and-water mix from the surface of the sea.

The numbers were impressive, but still woefully inadequate, with at least 1.5 million gallons of oil already released from BP's well, just based on the government's conservative estimates of the size of the leak. As Napolitano was giving her report, an oil slick about 130 miles long and 70 miles wide was creeping toward shore and threatening one of the most important wetland ecosystems in the world. The marshes of the Mississippi River delta had already been devastated by Hurricane Katrina and other Gulf storms in the previous five years, and a coating of petroleum could destroy the ability of the remaining wetlands to help reduce erosion and flooding along the four-hundred-mile Louisiana coast. One fisherman, charter operator Mark Stebly, reported on April 28 that oily gobs were already washing up on the Chandeleur Islands that stretch into the Gulf from the eastern tip of Louisiana.[5]

Coast Guard officials announced plans to conduct "controlled burns" to try to reduce the amount of oil flowing to the shoreline. Rear Admiral Mary Landry, now in command of the spill site, expressed hope that burns, which would be done miles from the coast, could reduce the amount of oil collected within fireproof booms by more than half, though she acknowledged that the downside was that a plume of soot and possibly toxic air pollutants would be spewed into the atmosphere.

The National Oceanic and Atmospheric Administration called that a reasonable trade-off in a "guide" to the burns the agency released to the public. "Based on our limited experience, birds and mammals are more capable of handling the risk of a local fire and temporary smoke plume than of handling the risk posed by a spreading oil slick," the NOAA said. "Birds flying in the plume can become disoriented, and could suffer toxic effects. This risk, however, is minimal when compared to oil coating and ingestion."

Cynthia Sarthou, director of a fifteen-year-old coalition of environmental groups called the Gulf Restoration Network, said the burns were objectionable but could be the lesser of several evils. "In the end, all of the options require us to choose potential harm to marine species in hopes of saving impacts to species in our coastal wetlands and beaches," she said.[6]

The first of what would become a long series of controlled burns was done late in the afternoon of April 28, not far from the site of the leak about fifty miles offshore. BP and the Coast Guard also began dumping tons of chemical dispersants on the spill, in hopes that some of the oil would break down before it reached the coast. Johnny Nunez, a charter boat operator who witnessed the aerial spraying, said that at times as many as two dozen planes at once were showering the Gulf with chemical agents. "It looked like World War II," he said.

By this time, President Obama was being briefed on a daily basis by Cabinet members involved in the response, but for the first week after the accident the White House kept an unusually low profile on the spill. Perhaps it was hoping BP would somehow be able to stop the leak before it became a full-blown crisis. When it was apparent the well was out of control, public concerns about what was happening in the Gulf started to mount, and some began questioning whether the White House was fully engaged in the crisis.

Obama had reason to be somewhat sheepish about the spill. He had unveiled a proposal on March 31 to expand offshore drilling in the Gulf, in the Arctic, and along other parts of the U.S. coast, and while he was advocating for more exploration he went out of his way to defend the safety record of the oil industry. "It turns out, by the way, that oil rigs today generally don't cause spills. They are technologically very advanced," Obama had said on April 2.

Now the president was faced with having to admit he was wrong and begin taking on an industry that was more accustomed to partnering with the government than being regulated by it. Obama finally made a forceful public statement on the spill during an event on education in the White House Rose Garden on April 28. "While BP is ultimately responsible for funding the cost of response and cleanup operations, my administration will continue to use every single available resource at our disposal, including potentially the Department of Defense to address the incident," he said.

The chief executive officer of BP was also on the defensive from the moment he received a 7:30 A.M. phone call during his breakfast the morning after the accident. Tony Hayward rushed to his London office for a briefing, but he had to know that the deaths of eleven workers and a massive blowout in the Gulf of Mexico would be as damaging to BP as the Texas City refinery disaster that killed fifteen people and injured dozens of others just five years earlier. "What the hell did we do to deserve this?" Hayward was reported to have asked his fellow executives at BP before he departed for the Gulf.[7]

During a stop at BP's U.S. headquarters in Houston and later on a tour through Louisiana and the Gulf, Hayward repeatedly tried to assure the American public that the company would do everything it could to contain and clean up all the spilled petroleum. "We take it with the utmost seriousness," he said in a written response to a question from *The New York Times*. "Nothing else matters right now."[8]

Hayward even evoked the wartime rhetoric of the late British Prime Minister Winston Churchill, who declared in a famous 1940 speech: "We shall go on to the end, we shall fight in France, we shall fight on the seas and oceans, we shall fight with growing confidence and growing strength in the air, we shall defend our island, whatever the cost may be, we shall fight on the beaches, we shall fight on the landing grounds, we shall fight in the fields and in the streets, we shall fight in the hills; we shall never surrender."

In an April 30 statement issued in a BP news release, Hayward said, "In the past few days I have seen the full extent of BP's global resources and capability being brought to bear on this problem, and welcome the offers of further assistance we have had from government agencies, oil companies, and members of the public to defend the shoreline and fight this spill. We are determined to succeed. We are determined to fight this spill on all fronts, in the deep waters of the Gulf, in the shallow waters and, should it be necessary, on the shore."

At the same time, BP officials tried to deflect some of the blame for the spill to its contractors on the drilling operation, especially the owner of the *Deepwater Horizon*, Transocean. "This accident took place on a rig owned, managed, and operated by Transocean," said BP public affairs chief Andrew Gowers at the end of April. "It involves the failure of a piece of equipment on that rig. So the unfolding events do not arise from a failure of BP's safety systems."

President Obama was having none of it. "BP is ultimately responsible for funding the cost of response and cleanup operations," he declared flatly on April 29.

Some analysts did give BP—and particularly Hayward—credit for trying to address the safety issues that had been exposed in Texas City and Alaska. But whatever progress had been made was blown away with the drilling rig on April 20. "It's a public relations disaster," Fadel Gheit, a managing director at Oppenheimer & Company, told the *Los Angeles Times* in late April. "Hayward has worked so hard to right the ship over the last three years, and now this. Everything he did is now meaningless."[9]

BP did acknowledge early in the crisis that it was not equipped to handle the disaster on its own, and it reached out for help not only from the government but to fishermen, contractors, and anyone in the public in a position to assist. BP specifically asked the Pentagon for help, particularly for underwater devices that might be more advanced than equipment that was available commercially. "It looks like they are trying to do everything they can," Phil Weiss, a senior analyst at Argus Research, told the *Los Angeles Times*.[10]

The biggest outreach effort was called Vessels of Opportunity, which offered to pay boat owners as much as several thousand dollars a day to take a crash course in cleanup methods and hit the waters to assist. The program attracted hundreds of fishermen left idle by restrictions on fishing in a large part of the Gulf—so many, in fact, that most times only about a third of the boats that were registered were called into cleanup service.

Eventually the program turned into its own public relations problem for BP. A contractor hired to train and assign boaters to the cleanup gave as many jobs to people from outside the region as it did to local fishermen affected by the spill, at one point leading a group of Louisiana and Mississippi fishermen to threaten a blockade of outsiders with vessels trying to get to the Gulf. "Why do you bring people from somewhere else to clean up your water?" the Reverend Jesse Jackson told a rally of shrimpers and other fishermen in Port Sulphur, Louisiana. "Make us the first choice for the cleanup of what someone else messed up."[11]

Many fishermen refused to work with BP, fearing their boats would be permanently tainted by oil from the cleanup work. Others were simply unwilling to accept money from the company whose actions had forced them into docks. Some went outside the restricted fishing zones and charged BP for the extra fuel needed on the longer trips.

Oystermen had no such options, as their beds in the shallow Gulf waters were the most vulnerable to damage from the spill.

"Oil has a real adverse effect on oyster reefs," longtime harvester Claude Duplessis of Pointe a la Hache in Plaquemines Parish told PBS's *NewsHour*. "In the reproduction stage, the oysters . . . put out a milk that's what we call spat. And this spat swims around in the water until it finds a clean, hard surface to attach itself to, and it grows from there. Now, if the oil coats the shell and the culch, then the spat can't stick, and . . . this can continue for years."[12]

Down the coast of the delta in Empire, Louisiana, Frank and Mitch Jurisich scrambled in late April to fill bags full of oysters even as the smell of oil wafted in from the sea; they feared it could be the last harvest for a family that had thrived on the delicacy for three generations.[13]

Louisiana, with 1.6 million acres of public oyster beds and hundreds of thousands of acres more of privately leased beds, is the heart of the nation's $1 billion oyster industry, and many worried that the spill would damage it for years. As the oil crept toward the shore, Louisiana Governor Bobby Jindal declared a state of emergency, making businesses eligible for state and federal aid, but if most of the beds were destroyed in 2010, they might not come back until at least 2013, industry leaders feared.

Most Louisiana shrimpers and hundreds of fishing charters and commercial fishermen were also knocked out of business just as a very promising season was about to begin. The sudden shift from anticipation to idleness fueled a sense of hopelessnesss in many fishing communities. "It's like a slow version of Katrina," said charter boat captain Bob Kenney, sitting at the dock in Venice, Louisiana, after most recreational fishermen canceled their plans, much as they did before the devastating hurricane moved up the Gulf in 2005.[14]

Plaquemines Parish president Billy Nungesser tried to give some hope to the despairing fishing communities by promoting plans to quickly build sand berms along the coast to block the incoming oil. "We're not doing everything we can do," Nungesser complained in late April, frustrated by the apparent lack of action by BP and the federal government.[15]

Nungesser, whose idea for the berms was backed by a fellow Louisiana Republican, Governor Jindal, was fast becoming the public face of anger in the Gulf, making frequent appearances in television interviews to express frustrations about the spill felt by many Americans.

He was the ideal man for the role. Nungesser was a fiercely independent, genuine entrepreneur who had dropped out of college to build a unique business turning shipping containers into portable living quarters for offshore oil workers. He lived inside one of the prototypes for three years to demonstrate its reliability, and had a compelling promotional video made showing a pipe dropping on to the roof of a fiberglass bunkhouse, and bursting through, and then showing the same test on his steel shelter bouncing right off. "Where would you rather sleep offshore tonight?" Nungesser asked the viewer as he stepped outside his container. Nungesser's business, General Marine Leasing, would reach sales of more than $20 million annually before he sold it in 2000 for a reported $18 million.[16]

A New Orleans native, Nungesser became familiar at an early age with Plaquemines Parish, which was essentially a long, narrow peninsula stretching southeast from New Orleans more than seventy miles into the Gulf. He often accompanied his father on the two-hour drive to Venice, at the southern end of the parish, to buy seafood for the family's canning company.

Nungesser came to love the parish, and his heart bled for it when Katrina roared through in 2005 and ravaged thousands of homes and businesses. He decided to follow in his father's footsteps—Billy Nungesser, Sr., had been chairman of the Louisiana Republican Party and was chief of staff to former governor Dave Treen, who in 1980 became the state's first GOP governor in more than a century. Nungesser, Jr., ran for parish president in 2006 and won—by a mere two hundred votes—on a platform of coastal restoration and infrastructure development.

A portly man with boundless energy, Nungesser made the BP oil spill a personal as well as a parish cause from the moment the *Deepwater Horizon* exploded. In a feature about Nungesser on CNN, his friend Rene Cross described him as a nonstop crusader. "I don't think Billy ever really relaxes," Cross said. "I can get him out on the boat or hunting, and he'll take it easy for a bit, but within ten minutes, it's back to the business of Plaquemines Parish. But he loves it. He takes great joy and pride in contributing to the improvement of the community."

While Nungesser was developing plans for a sand berm to protect his parish, environmentalists in the Gulf region were anxiously watching reports of the spill's movement as it threatened a host of species in the lush coastal wetlands. "We are in a breathless waiting game," Melanie Driscoll of the National Audubon Society told the *New Orleans Times-Picayune* in late April. "The timing is really quite bad. For some species, it's the beginning of the nesting season, and there are birds on nests and eggs on the ground, on barrier islands, and beaches where the oil is likely to occur."[17]

Spring is also migration time for millions of shorebirds that rely on the wetlands for food and nesting; there were fears that many would descend to their usual feeding grounds for fish only to be covered by oil. Many would simply disappear to the bottom of the sea, feared Kerry St. Pe, program director of the Barataria-Terrebonne National Estuary Program who formerly worked with the Louisiana Department of Environmental Quality on oil-spill cleanups. "Shrimp die and crabs die and oysters die, but they don't float to the top," St. Pe said. "You just never see them, but the damage is often severe."[18]

The tourism industry went into a state of shock, as the white sands of the Gulf beaches were threatened at the very start of the warm weather. The approaching slick frightened prospective vacationers, and thousands canceled trips to the Gulf.

A group from the National Wildlife Federation took an aerial tour of the Gulf in early May and Jeremy Symons said it looked "as if somebody just took a paintbrush and just put varnish atop the ocean."[19]

BP was now spending at a rate of $6 million a day on efforts to try to contain the oil and cap the well. The company was also being hit financially by a precipitous drop in its stock prices, which fell 13 percent in the last ten days of April and wiped $20 billion off BP's market value. But while it was taking those hits, the company quietly reported profits of $6.1 billion in the first quarter of 2010—an increase of $2.6 billion from the first quarter of 2009. Revenue increased more than 50 percent, to $74.42 billion, from the same quarter of the previous year.

The entire oil industry was slammed at the end of April when President Obama imposed a moratorium on new offshore drilling until an "adequate review" was done of the BP accident. Fears mounted that with Louisiana's second-biggest industry already reeling from the restrictions on fishing, a shutdown of its biggest industry—oil and gas—could be devastating to the state's economy, especially in the depths of a recession.

"I continue to believe that domestic oil production is an important part of our overall strategy for energy security," Obama told reporters in the White House Rose Garden, "but I've always said it must be done responsibly, for the safety of our workers and our environment."

Interior Secretary Salazar stressed that production could continue at the 30,000 wells already in the Gulf, including several hundred in the deeper waters. Only new wells would be put on hold, he said, enabling most of the 35,000 workers on more than a thousand rigs and platforms to continue the critical job of producing 1.7 million barrels of oil per day.

Others called for even greater restrictions on the oil industry. All six U.S. senators from the three states on the Pacific Coast proposed a permanent ban on drilling there, even though one was already in place. A plan for moving toward limited drilling off the coast of California, suggested before the BP accident by Governor Arnold Schwarzenegger, was discreetly withdrawn.

The Obama administration increased pressure on BP to find a

way to plug the leak. "Our job basically is to keep the boot on the neck of British Petroleum to carry out the responsibilities they have both under the law and contractually to move forward and stop this spill," Salazar said on CNN's *State of the Union* on Sunday, May 2.

Later that day, Obama also made his first visit to the Gulf since the spill began. On a rainy afternoon he met with a group of fishermen, now faced with the prospect that Gulf waters from the mouth of the Mississippi River east to Pensacola, Florida, would be placed off-limits. The president offered assurances that the government would insist on a full cleanup at no cost to taxpayers. "BP is responsible for this leak. BP will be paying the bill," he said.

U.S. Attorney General Eric Holder went to the Gulf a few days later to reinforce the message. Holder said the Justice Department was investigating whether there was any "misfeasance" or "malfeasance" by companies involved in the spill, although he said the government's first priority was stopping the leak.

BP's Tony Hayward responded in an interview with *The Wall Street Journal*. "We will fix this," he said, previewing ads the company was preparing to unleash in the coming weeks.[20]

But hundreds of lawsuits were already piling up against the company. Louisiana shrimpers were among the first to file for damages of millions of dollars, charging that BP was destroying their livelihoods. Hoping to ward off some litigation, Hayward repeatedly stated publicly that the company would pay all "legitimate claims for business interruption."

BP also moved to try to limit its exposure to lawsuits by appealing directly to potential litigants. Several newspapers reported that fishermen in the Gulf region were being asked to sign waivers of damage claims in return for a $5,000 payment. Alabama Attorney General Troy King ordered the company to stop circulating such offers in his state, though BP denied it was doing so. BP spokesman Daren Beaudo told the *Mobile Press-Register*: "To the best of my knowledge BP did not ask residents of Alabama to waive their legal rights in the way that has been described."[21]

BP did agree to change the wording of waivers it had asked workers to sign before they were hired to help with the cleanup, after one fisherman, George Barisich, went to federal court to argue the agreement compromised his constitutional rights. A copy of the waiver leaked to the press stated: "I hereby agree on behalf of myself and my representatives, to hold harmless and indemnify, and to release, waive, and forever discharge BP Exploration and Production Inc., its subsidiaries, affiliates, officers, directors, regular employees and independent contractors." When a federal judge ruled that the language was too broad, BP agreed to rewrite it and said it would disregard all forms that had already been signed.[22]

As the number of lawsuits continued to climb within a few weeks of the accident, BP asked a special multidistrict litigation panel to order all the cases to be heard in Houston, where BP had its headquarters. The U.S. Judicial Panel on Multidistrict Litigation said it would decide on the request at its July session in Boise, Idaho.

One case, separate from the spill litigation, demonstrated a growing distrust of BP's drilling operations in the Gulf. A group called Food & Water Watch, after obtaining BP documents from a former oil worker turned whistle-blower, asked the U.S. District Court in Houston to order BP to stop production on its *Atlantis* drilling platform because there was evidence it was being operated unsafely. The whistle-blower, former BP contract worker Kenneth Abbott, charged that the rig had none of the required documentation on board describing the structure and design of underwater equipment, so that in effect the *Atlantis* was being operated without an owner's manual. "A similar spill from the BP *Atlantis* Facility would have a calamitous effect on marine life, endangering the public health and the economic viability of plaintiffs," the complaint said.[23]

BP denied the charges, even though its own internal ombudsman, former federal judge Stanley Sporkin, said he had verified the allegations, but the mere fact that it had to fight the legal battle showed how much the company was under attack on numerous fronts.

The wave of litigation against BP had CEO Hayward firmly on the defensive, proclaiming on a CBS news show that the company was not liable for the spill from the Macondo well. "This is not our accident, but it's our responsibility," he said. Hayward was attempting to deflect responsibility to Transocean, which owned the *Deepwater Horizon* rig that BP was leasing when the April 20 explosion occurred. "It was their rig and their equipment that failed, run by their people with their processes," Hayward said. "But our responsibility is the oil, and the responsibility is ours to clean it up. And that's what we're doing."

Hayward also went to Washington to try to reassure angry lawmakers on Capitol Hill. He met privately with key members to brief them on the company's plan to lower a giant containment dome on the leaking well in hopes of capturing much of the oil.

BP lowered its dome over the well May 8, but immediately ran into problems. The opening at the top filled with icy solids in the high-pressure, low-temperature conditions a mile below the surface, so none of the oil captured in the box could be pumped out. It had been hoped that as much as 85 percent of the oil spewing from the well could be recovered, but now all BP had was a ninety-eight-ton steel structure that had taken weeks to build and deploy sitting uselessly on the bottom of the ocean.

BP tried desperately to put the best face on the setback. "I wouldn't say it has failed," Doug Suttles, BP's chief operating officer, said in a conference call with reporters. "What I would say is what we attempted to do last night wasn't successful."

Now the company would try plan B, described by Suttles as putting a "top hat" on the well, a device smaller than the containment dome but with a tighter fit. "We're going to pursue the first option that's available to us and we think it'll be the top hat," Suttles said.

There were other plans in the works if the top hat failed, BP officials said. One was to use a smaller containment box, with steps taken to prevent freezing at the opening on top. Another idea was to cut through the leaking well pipe with a saw to put a flat, smooth surface on the top that could be cleanly capped. Plans were also being developed for a "top kill," which would involve pouring concrete and debris into the blowout preventer that still sat atop the well, in hopes of clogging it up to stop the flow of oil and gas.

And then there was a final, more certain plan to kill the well, but it was one that would take months to complete. A relief well could be drilled into the reservoir to seal the leaking well from below, but drillers who had already started the process had more than 13,000 feet of rocks and undersea formations to cut through. At best, the relief well would not be completed until August, and that was only if there were no delays from the Gulf's sometimes active hurricane season, which was about to begin in June.

The appearance that BP was flying by the seat of its pants to stop the leak brought cries from politicians at the state and federal levels to demand that the government take over the entire spill response. Many blamed BP for being completely unprepared, and some experts in the energy industry had to agree. "The only thing that's clear is that there was a catastrophic failure of risk management," Nansen Saleri, a oil expert formerly with Saudi Arabia's state-owned petroleum company, Saudi Aramco, told *The Wall Street Journal*.[24]

It should have been no surprise to regulators in Washington that BP was scrambling for a solution. It was becoming apparent that the company had done little to prepare for worst-case scenarios in its growing deepwater exploration program, even though federal law required extensive response plans to be prepared as a condition for obtaining drilling permits. Journalists who dug up BP's "Regional Oil Spill Response Plan—Gulf of Mexico," approved by the government in July 2009, found no mention in its 582 pages of how a major blowout would be handled.

Congressional investigators who put BP's plan and others like it under the microscope discovered that many sections appeared to be part of a generic, cookie-cutter document that had been prepared for multiple companies that filed versions with the federal Minerals Management Service. It turned out that at least five of the response plans were all written by the Response Group, a Houston company with offices in six other cities, and all were equally deficient.

"It could be said that BP is the one bad apple in the bunch," said

Representative Bart Stupak (D-Michigan), at a May hearing of his congressional investigations subcommittee. "But unfortunately they appear to have plenty of company. Exxon and the other oil companies are just as unprepared to respond to a major oil spill in the Gulf as BP."

The most galling thing about the plans—aside from the fact that three of them provided contact information for a Florida scientist who had been dead for five years—was the section titled "Sensitive Biological & Human-Use Resources" that listed "seals, sea otters, and walruses" as animals that could be harmed by an oil spill in the Gulf. All are cold-climate species that "have not called the Gulf of Mexico home for three million years," noted Representative Edward Markey (D-Massachusetts) at the hearing on BP's response capabilities.

ExxonMobil CEO Rex Tillerson admitted at the hearing that his industry's preparedness plans were "an embarrassment" and when asked how the companies expected to respond to major spills, he conceded, "We are not well equipped to handle them."

"These oil spill response plans suffer from what I would consider a 'failure of imagination,'" said Representative Nick Rahall (D-West Virginia) at a separate hearing of the House Natural Resources Committee that he chaired. "It seems to me that there should be a plan B, C, and D in place before the accident occurs, not created in haste while millions of gallons of oil are spewing into the Gulf."

BP spokesman Steve Rinehart responded that the company had to improvise because of the "unforeseen circumstances" of a deepwater leak and a failure of the blowout preventer to stop it. "Nobody foresaw an incident in which something like this occurred," he said.

Hammond Eve, who spent years working on environmental impact studies for offshore drilling at the Minerals Management Service, gave BP the benefit of the doubt because a huge leak so far below the surface was never anticipated by either the industry or the government. "We never imagined that it would happen because the safety measures were supposed to work and prevent it from happening," Eve told *The Washington Post*. All the experts agreed that if a deep-sea blowout occurred, "it would be shut down fairly soon and a discrete amount of oil would be released and these cleanup measures would begin and you would never end up with a situation like this," he said.[25]

Coast Guard Commandant Thad Allen also offered some sympathy for BP in the same *Post* article on May 1. "We're breaking new ground here," Allen said. "It's hard to write a plan for a catastrophic event that has no precedent, which is what this was."

BP also defended its actions. "You have here an unprecedented event—never before have you seen a blowout at such depth and never before has a blowout preventer failed in this way," said spokesman Andrew Gowers. "The unthinkable has become thinkable, and the whole industry will be asking searching questions of itself."

Members of Congress were asking probing questions of the whole industry in hearings held three weeks after the *Deepwater Horizon* sank, and many were infuriated by the answers they were given. Top executives from BP, Transocean, and Halliburton sat together at the witness table but each one laid the blame for the accident on the others.

"Transocean's blowout preventer failed to operate," testified Lamar McKay, chairman of BP America Inc. And it was Transocean,

McKay said, that "had responsibility for the safety of the drilling operations."

Transocean CEO Steven Newman questioned the quality of the work done by Halliburton, which was doing the cement job to seal the well when oil and gas broke through. "Were all appropriate tests run on the cement and the casing?" Newman asked.

Tim Probert, chief of health, safety, and environment for Halliburton, said that every step taken by his company on the rig was done "as directed by the well owner," BP.

An exasperated Senator Lisa Murkowski of Alaska, the ranking Republican on the Senate Energy and Natural Resources Committee and a fervent supporter of the oil industry, lectured the executives as if they were children in a grade school classroom. "I would suggest to all three of you that we are all in this together," she said. If the Gulf spill led to new limits on drilling, she said, "not only will BP not be out there, but the Transoceans won't be out there to drill, and the Halliburtons won't be out there cementing."

President Obama fumed about the blame-shifting a few days later. "I did not appreciate what I consider to be a ridiculous spectacle during congressional hearings into this matter," he said. "You had executives of BP and Transocean and Halliburton falling over each other to point the finger of blame at somebody else. The American people cannot have been impressed with that display, and I certainly wasn't.

"I understand that there are legal and financial issues involved, and a full investigation will tell us exactly what happened," Obama continued. "But it is pretty clear that the system failed, and it failed badly. And for that, there's enough responsibility to go around.

And all parties should be willing to accept it. That includes, by the way, the federal government. I will not tolerate more finger-pointing or irresponsibility."

The president tried to match his words with action, sending Interior Secretary Salazar and Energy Secretary Steven Chu to the Gulf to step up pressure on BP. Salazar announced the formation of a new oversight board, and Chu said government supercomputers would be used to get better assessments of what was happening at the bottom of the Gulf.

The scene in the Capitol hearing rooms was perfect fodder for NBC's *Saturday Night Live* the following weekend, on May 15. The opening skit featured actors playing Transocean's Newman, Halliburton's Probert, and BP CEO Hayward standing at a lectern as if giving a press briefing. The faux Hayward opened by describing all the company's plans for closing down the well—the "top hat," the "giant tube," and the "junk shot"—but added that there was more on the drawing board. "Those are the only plans we've announced so far, but, tonight, we've come together to assure you we have *many* other ideas. Ideas formulated by our top scientists, using state-of-the-art technology. The first plan is called . . . 'dolphins with mops.'"

"That's where we round up a bunch of dolphins and Scotch-tape mops to their fins," says the actor playing Newman. "It may not work, but, rest assured, Halliburton *will* make a profit!" pipes in the Probert character.

Hayward concludes: "No matter what happens, we just want to assure the American people that we *will* stop this leak, we *will* clean up this mess, and we *will* get back to doing what we do best—robbing you blind at the gas pump!"

While the show was airing, BP and its contractors were trying a new method of capturing oil from the well that had just been cooked up by their engineers, who were now calling the well "the beast" that could not be tamed. The idea was to place a tube surrounded by rubber flaps into the gushing well to funnel off at least some of the oil that was being released into the sea.

The first attempt to insert the siphon failed when the tube dislodged, but a second effort was successful around midnight on Saturday, May 15. "It's working as planned," said BP executive Kent Wells at a Sunday afternoon press briefing in Houston. "So we do have oil and gas coming to the ship now, we do have a flare burning off the gas, and we have the oil that's coming to the ship going to our surge tank." The company later estimated it was pulling in a mixture of oil, gas, and water at a rate of about a thousand barrels a day, or 42,000 gallons. It was hoped that the amount could be doubled soon. It was only a fifth of the oil estimated to be flowing from the well, but it was the first good news from the Gulf in nearly a full month.

The next step, according to Doug Suttles, would be the "top kill," an attempt to pump about 50,000 barrels of gunk into the well at high pressure, in hopes of countering the force coming out. Then the well could be sealed with mud and cement, stopping the flow until the relief well could be drilled. The plan was to start the process by the weekend of May 22 and 23.

During a flight over the site with Louisiana Governor Jindal, Suttles also made a pledge that once the well was sealed, it would never be reopened to pump out oil. "There is absolutely no intent to ever, ever produce this well," he said. "We would pump cement into it and close it."[26]

Amid the signs of progress, Hayward again attempted to downplay the crisis. "The Gulf of Mexico is a very big ocean," he told a reporter from London's *Guardian* newspaper in an interview at the BP command center in Houston. "The amount of volume of oil and dispersant we are putting into it is tiny in relation to the total water volume."[27]

Hayward also told Britain's Sky News a few days later that the impacts of the spill would be minimal. "I think the environmental impact of this disaster is likely to have been very, very modest," he said. "It is impossible to say and we will mount, as part of the aftermath, a very detailed environmental assessment, but everything we can see at the moment suggests that the overall environmental impact will be very, very modest."[28]

At the same time, the company seemed to be retreating from its earlier appeals for help from anyone who could assist. *The New York Times* reported on May 15 that BP had rejected requests by scientists to lower high-tech devices to the wellhead to try to better gauge the amount of oil being released. "The answer is no to that," said a BP spokesman, Tom Mueller. "We're not going to take any extra efforts now to calculate flow there at this point. It's not relevant to the response effort, and it might even detract from the response effort."[29]

Critics said the company was trying to cover up the true extent of the spill to reduce its legal exposure, since future penalties would likely be based on the amount of oil released into the environment in violation of federal laws. "If they put off measuring, then it's going to be a battle of dueling experts after the fact trying to extrapolate how much spilled after it has all sunk or has been carried away," said Lloyd Benton Miller, a lawyer involved in litigation that resulted from the *Exxon Valdez* spill. "The ability to measure how much oil was released will be impossible."[30]

Even as BP tried to minimize the impacts of the spill, scientists doing research away from the well site were beginning to collect evidence of massive plumes of oil stretching for miles below the surface. "This monster's turned invisible," commented a frustrated Billy Nungesser on land in Plaquemines Parish. "How do you fight that monster when it's invisible?"[31]

Some scientists speculated that a major reason for the undersea plumes was the extensive use of chemical dispersants that broke up hydrocarbons and disrupted the usual principle of physics that oil floats on water. Many experts feared that the chemicals were altering or killing plankton that was the critical first link in the sea's food chain, or possibly causing toxic contaminants to settle on the ocean floor, wiping out sensitive coral or poisoning bottom-feeding species.

At least a half-million gallons of dispersants had been dumped on the spill in the first month after the blowout, reports indicated, and the chemical warfare did seem to reduce the amount of heavy oil floating toward the beaches and wetlands. But there were many questions about the unseen effects.

At one point the Environmental Protection Agency ordered BP to start using a less toxic form of dispersant than the one most commonly being used in the Gulf, Corexit. The chemical mixture in Corexit was known to be toxic to humans and had been banned in a number of other countries, including Britain.

BP argued that the chemical was the most effective at breaking up oil, and would break down fairly easily in water. And even if it was absorbed by fish or other ocean creatures, most would quickly eliminate it without harm to their systems. "It appears that the application of the subsea dispersant is actually working," BP's Doug Suttles said on May 15. "The oil in the immediate vicinity of the well and the ships and rigs working in the area is diminished from previous observations."

That didn't answer the questions about what the dispersants were doing to the oil and where it was ending up. "By dispersing the stuff at depth, it creates essentially smaller globules of oil [and] it makes the oil more likely to be affected by even slow-moving currents," said James H. Cowan, Jr., a professor of oceanography and coastal sciences at Louisiana State University. "We just don't know where it is, and we don't know where it's going."[32]

The Coast Guard's on-scene commander, Rear Admiral Mary Landry, agreed there were concerns about the chemicals but said federal scientists approved their use. "That threshold we crossed was not done lightly," Landry said. "It was done in cooperation with all the federal agencies. . . . It's all a series of trade-offs. We are really trying to minimize the environmental impact of all these methods being done."

BP Managing Director Robert Dudley announced on May 21 that the company would put up $500 million to research the effects of the dispersants in the Gulf, including how they move and what impacts they have on marine life. "We will be studying this for many, many years to come," he told Judy Woodruff on the *PBS NewsHour*.[33]

Representative Edward Markey (D-Massachusetts), the chairman of a House committee on global warming, begged EPA administrator Lisa Jackson to be very watchful of the company's actions and their effects. "The release of hundreds of thousands of gallons of chemicals into the Gulf of Mexico could be an unprecedented, large, and aggressive experiment on our oceans, and requires careful oversight by the Environmental Protection Agency and other appropriate federal agencies," Markey wrote to Jackson on May 17.

Markey also pressed BP hard for more information about the leak, and when he learned the company had a deep-sea camera focused on the gusher, he demanded the link and posted it on his committee's Web site. Almost immediately the live video feed was a staple of televised reports about the spill, further enraging the public about the inability to stop it.

The images of the dark liquids spewing uncontrolled into the Gulf added to the national sense of urgency about the environmental crisis, and led to more calls for federal authorities to take over the response. "The government is in a situation where it's required to be in charge," William Funk, a former Justice Department attorney, told *Rolling Stone* magazine. An unnamed scientist also was quoted in the article as saying that keeping BP in charge of the spill response was "like a drunk driver getting into a car wreck and then helping the police with the accident investigation."[34]

But a government takeover might not have been much of a solution. While BP was the favorite whipping boy in Washington, congressional committees also were bearing down on the Minerals Management Service for its apparent failure to keep the industry under control. President Obama promised to break the agency apart and start over.

On May 11, the administration announced it would split the

MMS into two divisions, one to regulate drilling and one to manage leases and collect royalties. This would eliminate the agency's apparent conflict of interest and let "the American people know they have a strong and independent organization holding energy companies accountable," Interior Secretary Salazar said in a statement. A third division was added to the new MMS plan later—a unit to oversee the safety of workers in the oil and gas industries. The new offices would be called the Bureau of Ocean Energy Management, the Bureau of Safety and Environmental Enforcement, and the Office of Natural Resources Revenue.

Within a week of the announcement, a reshuffling began at the MMS. The veteran manager of offshore drilling programs, Chris Oynes, said he would retire at the end of May. Oynes had been in hot water with some members of Congress for several years after he was criticized for signing off on royalty exemptions for hundreds of offshore leases, costing the government millions of dollars in revenues.

Two weeks later, the head of the MMS, S. Elizabeth Birnbaum, announced her resignation. "I'm hopeful that the reforms that the Secretary and the Administration are undertaking will resolve the flaws in the current system that I inherited," Birnbaum said in a letter to Salazar. She had been in the job for less than a year, an environmental lawyer tapped by Salazar to clean up an agency that had been plagued by scandal in the previous administration. While she was considered smart and tough, her lack of experience with the energy industry made her job much more difficult.

The depth of the problems at the MMS became apparent in a new report by the Interior Department inspector general released in late May.

Employees directly responsible for drilling oversight in the Gulf of Mexico often accepted gifts from oil companies and shared a prevailing attitude that they were part of the industry, the IG found. "We discovered that the individuals involved in the fraternizing and gift exchange—both government and industry—have often known one another since childhood," wrote the report's author, Acting Inspector General Mary Kendall. An MMS district manager in Louisiana, Larry Williamson, told IG investigators that there were many bonds between industry workers and their regulators. "Obviously, we're all oil industry," he said. "Almost all of our inspectors have worked for oil companies out on these same platforms. They grew up in the same towns. Some of these people, they've been friends with all their life. They've been with these people since they were kids. They've hunted together. They fish together. They skeet shoot together. . . . They do this all the time."

There was also evidence of a lax working environment within the agency, including some employees who regularly watched pornography on their desktop computers, and two employees who admitted to using illegal drugs at work. Another worker told investigators that MMS inspectors would sometimes allow oil company employees to fill out the government inspection forms in pencil and the MMS employees would fill them in with ink later before submitting them to headquarters.

The findings were presented to U.S. attorneys in Louisiana for possible charges, but no action was taken. However, the report made clear that the culture at the MMS badly needed reform, and it would take a tough manager to accomplish it. President Obama decided in June to appoint a former federal prosecutor and inspector general for the Justice Department, Michael Bromwich, to take on the assignment.

By the end of May there were more reports from scientists finding underwater oil plumes, some more than a half mile beneath the surface and stretching more than twenty miles in length. The fact that they were documented by researchers from multiple institutions—including the University of South Florida, the University of Georgia, and Southern Mississippi University—gave added credibility to the findings and seemed to provide a reasonable explanation for why there was not more oil floating on the Gulf surface, considering the rate that it was pouring from the well.

But Hayward dismissed the plumes as a myth. "The oil is on the surface," the BP leader insisted. "Oil has a specific gravity that's about half that of water. It wants to get to the surface because of the difference in specific gravity."[35]

Countered marine scientist James Cowan of Louisiana State University: "There's been enough evidence from enough different sources." Cowan himself reported finding an oil plume about fifty miles from the spill site that was at least four hundred feet below the surface.[36]

"BP is burying its head in the sand on these underwater threats," Congressman Markey said in a written statement after the scientists' findings were reported in the press. "These huge plumes of oil are like hidden mushroom clouds that indicate a larger spill than originally thought and portend more dangerous long-term fallout for the Gulf of Mexico's wildlife and economy."

With BP and the government appearing to be stumbling, offers of assistance poured in from experts around the world, as well as from ordinary citizens with ideas for how to stop the leak and clean up the oil. Two of the most publicized suggestions came from Hollywood celebrities—actor Kevin Costner and movie director James Cameron.

Costner revealed through his partners that he had invested millions of dollars in "centrifugal oil separators" after taking an interest in maritime issues while making the film *Waterworld* that was released in 1995. He offered to have them brought to the Gulf for use in the cleanup, and BP took some initial interest. "BP has agreed to test Mr. Costner's machines," said BP spokesman Mark Proegler. "Of course, they need to meet regulations with respect to discharge."[37]

Costner's offer raised hopes along the coast, and Plaquemine Parish's Nungesser was quick to suggest that the machines be tried there first to stem the tide of oil that was damaging fragile wetlands. "We've already lost twenty-four miles of marshland," Nungesser said. "Everything in it—frogs, crickets, fish, and plant life—is dead and never coming back."

Cameron, meanwhile, offered to let BP use underwater equipment that he had purchased to film the hit movies *Titanic* and *Avatar.* BP wasn't as receptive to Cameron's proposal, and the Oscar-winning director lashed out angrily. "Over the last few weeks, I've watched, as we all have, with growing horror and heartache, watching what's happening in the Gulf and thinking those morons don't know what they're doing," Cameron said at a digital conference on June 2. He later backtracked, telling talk show host Larry King on CNN that his remarks were taken out of context. "They're not morons," Cameron said. "There are good engineers out there."[38]

Thousands of other ideas for dealing with the disaster were phoned in to a BP hotline or brought directly to the Gulf by aggressive entrepreneurs. In fact, *USA Today* reported later in the year that BP had received about 123,000 ideas, including 80,000 for plugging the well and 43,000 for cleaning up the oil. Many were not feasible, such as the use of a nuclear bomb to destroy the well. "I had fifty nutcases a day walk through the door," David Kinnaird, a BP project manager, told the newspaper.[39]

But some suggestions showed promise. Kinnaird checked out an invention from a Massachusetts businessman, Scott Smith, for a type of foam that would absorb oil but not water. "This was completely different from anything I'd seen," Kinnaird said, and BP ended up buying 2 million square feet of the foam, called Opflex.

Not everyone at BP was impressed by the public outpouring. "We didn't miss anything great," said Hunter Rowe, a BP manager who sifted through suggestions. But some said that both BP and federal officials did miss opportunities because they were not open to new ideas and were often hung up on legal issues. Dwayne Spradlin, the CEO for a digital company called InnoCentive that promotes sharing of technological and scientific solutions, said many good suggestions were rejected. "The notion that we weren't using all the tools we had is incredible to me," Spradlin told *USA Today*. "The world deserves better."

The entire world was focused in early summer on the effects of the spill, which were becoming more serious every day. Photos of oil-covered birds and turtles dominated the news and were displayed on many Web sites. Scenes of cleanup workers wearing protective suits on Gulf beaches were scaring away thousands of visitors who usually flock there as soon as the weather warms up. And anger and frustration was growing across the nation as the Gulf crisis officially surpassed the *Exxon Valdez* accident as the worst oil spill in U.S. history. A new government estimate at the end of May put the flow from the well at between 12,000 to 19,000 barrels a day, well beyond the 5,000-barrels-a-day estimate that had been made a month earlier by the National Oceanic and Atmospheric Administration.

President Obama appointed an independent panel to determine once and for all how much oil was being released into the Gulf, with no representative of BP allowed to participate. The president also appointed a bipartisan commission to recommend changes in the government's offshore drilling program. A separate panel led by the Coast Guard and the MMS was investigating the causes of the *Deepwater Horizon* accident.

Markey, through his Select Committee on Energy Independence and Global Warming, released documents he obtained from BP showing the company had known for a month that the well was spewing more than 14,000 barrels a day, yet was still insisting publicly that the flow was less than a third that rate.

Oil sightings were becoming more frequent onshore. The NOAA reported that more than sixty-five miles of Louisiana coast had now been stained by oil.

Louisiana Governor Jindal said at a press conference exactly one month after the accident that it was not just an oil slick washing into the wetlands. "These are not tar balls, this is not sheen," said Jindal at a news conference in Venice, Louisiana. "This is heavy oil we are seeing in our wetlands."

People were turning to the Almighty for help: ABC's *World News* showed a sign outside a Louisiana church that read, PRAY FOR A SOLUTION TO THE OIL LEAK PROBLEM. Inside the Hosanna Church in Marrero, news anchor Diane Sawyer met with fishermen who described how they feared an end to the only livelihood they knew. "There's nothing out, there's not boats, no booms, nothing stopping it [the oil] from coming in," said Bonnie Gross, whose husband is a commercial fisherman. "So now, the areas that they could fish, they can't."[40]

Even far from the spill, Americans were venting their outrage. An online movement began in late May calling for a boycott of BP gas stations, just as the peak summer driving season was getting started. Analysts said a boycott would do little to hurt BP, which had $240 billion in sales worldwide in 2009, but would only harm independent station owners who simply carried the BP brand. Some of those owners openly considered reverting to their earlier brand names like Amoco, to avoid a backlash from consumers.

Politicians used BP as a daily punching bag, too, and Republicans turned the unending crisis into regular attacks on the Democrats in charge of both the White House and Congress. Former Alaska Governor Sarah Palin, who made the "drill, baby, drill" slogan a battle cry as the Republican vice presidential candidate in 2008, offered her support for continued offshore drilling, but couldn't resist a few shots at the Obama administration on the Fox News Channel.[41]

"I don't know why the question isn't asked by the mainstream media and by others if there's any connection with the contributions made to President Obama and his administration and the support by the oil companies to the administration," Palin said in her famous rambling manner. "If there's any connection there to President Obama taking so doggone long to get in there, to dive in there, and grasp the complexity and the potential tragedy that we are seeing here in the Gulf of Mexico—now, if this was President Bush or if this were a Republican in office who hadn't received as much support even as President Obama has from BP and other oil companies, you know the mainstream media would be all over his case in terms of asking questions why the administration didn't get in there, didn't get in there and make sure that the regulatory agencies were doing what they were doing with the oversight to make sure that things like this don't happen.

"There is certainly a double standard," Palin added, "and while there is plenty of blame to go around, let's not let the Obama administration off the hook as well. While certainly Republicans have received money from oil companies, out of the last twenty years BP has given President Obama the most."

The last statement might have been a reach, *Washington Post* columnist Ruth Marcus pointed out after Palin's appearance on Fox. BP employees did give $71,051 to Obama during the 2008 presidential campaign and about half that amount to his opponent, Senator John McCain (R-Arizona), Marcus reported. But overall the McCain-Palin campaign received $2.4 million from oil and gas interests, and the Obama-Biden campaign collected only $900,000 from the industry, she said, citing data from the Center for Responsive Politics.[42]

244] POISONED LEGACY

There were other attacks on the Obama administration by GOP leaders. Louisiana's Republican governor, Jindal, said on ABC's Sunday talk show *This Week* at the end of May that "I think there could have been a greater sense of urgency" in the government's response. Senator David Vitter (R-Louisiana), chimed in on CNN, saying, "There has been failure, particularly with the effort to protect our coast and our marsh. . . . The state and locals came up with a plan on emergency dredging barrier islands well over two weeks ago. For over two weeks, the [Army Corps of Engineers] and other federal agencies dragged their feet. Then, they approved moving forward with 2 percent of that plan."[43]

The top environmental adviser in the White House, former EPA administrator Carol Browner, defended the administration on NBC's *Meet the Press.* "The president has been in control from the beginning," she said. "He knows what's going on. He's in charge . . . I'm the point person in the White House. But we also have our Cabinet fully deployed—everybody from Homeland Security, to DOD, to the EPA, to Interior, to NOAA. This is—as I said—this is a very, very large disaster. And we are deploying all of our resources."[44]

Obama himself responded with a show of action, after clarifying that he took full responsibility for solving the crisis. "I'm the president and the buck stops with me," he said. With that, he announced that exploratory drilling off the Alaska coast would also be suspended, as new drilling in the Gulf had been halted earlier.

The expansion of the ban on new drilling angered Alaska lawmakers, even some of Obama's fellow Democrats. They noted that the move would delay five promising wells that Shell Oil Company had planned to start drilling in the Chukchi and Beaufort seas off the Alaska coast in June or July. The MMS estimated there could be as much as 19 billion barrels of oil beneath the two bodies of water.

"The Gulf of Mexico tragedy has highlighted the need for much stronger oversight and accountability of oil companies working off-shore, but Shell has updated its plans at the administration's request and made significant investments to address the concerns raised by the Gulf spill," said Senator Mark Begich (D-Alaska). The state's Republican senator, Lisa Murkowski, was also critical. "If the delay is for a season to ensure we have the highest levels of protection in place, that's one thing. But if it means that existing permits are al-lowed to lapse—effectively killing Shell's participation in Alaska—that's not acceptable to me or Alaska." Murkowski said she had talked with Shell's president, Marvin Odum, and been told that the company might have to suspend its plans if the moratorium lasted into 2011.

The White House pushed even harder on BP in the Gulf, de-manding that the company begin sharing data on the spill more openly with the government and the public. Homeland Security Secretary Janet Napolitano and EPA administrator Lisa Jackson wrote to CEO Hayward asking for "any and all sampling and/or monitoring plans, records, video, reports collected by BP, its con-tractors, subcontractors, agents, or employees," as well as any "re-ports of internal investigations." Napolitano and Jackson said they wanted all the information posted on the Internet, arguing that the American public is "entitled to nothing less than complete trans-parency."

The investigative news outlet ProPublica reported in late May that the EPA also was considering "debarment" of BP from future federal contracts, a drastic step that could cost the company billions of dollars and push it out of the Gulf. Few environmental attorneys believed the government would take that step against a company that provided much of the fuel used by the U.S. military, but the threat was a clear sign that top government officials were losing their patience with BP.[45]

"I am angry and frustrated that BP has been unable to stop the leak," Interior Secretary Salazar said after more than a month of failures to cap the well. "We're thirty-three days in, and deadline after deadline has been missed."

Despite the efforts to keep the focus on BP, talking heads on nonstop cable news networks increasingly argued that the Gulf spill could be for Obama what Hurricane Katrina was for his predecessor, George W. Bush—a total breakdown of government response. (Some weren't sure that was an apt comparison, though: "Katrina was nothing but rain, water, and wind. This is poison. It's gas," oysterman Arthur Etienne said on CNN.)[46]

Asked at a May 28 news conference if he felt he had made mistakes in responding to the crisis, Obama responded, "Where I was wrong was in my belief that the oil companies had their act together when it came to worst-case scenarios." But when he was told some were referring to the Gulf spill as "Obama's Katrina," the president bristled. "I'll leave it to you guys to make those comparisons, and make judgments on it, because what I'm spending my time thinking about is how we solve the problem." In the end, he said, "I'm confident that people are going to look back and say that this administration was on top of what was an unprecedented crisis."

Obama also offered a personal comment, saying he was "angry and frustrated" by the ongoing crisis. "And it's not just me, by the way," he said. "When I woke this morning and I'm shaving and Malia knocks on my bathroom door and she peeks in her head and she says, 'Did you plug the hole yet, Daddy?' Because I think everybody understands that when we are fouling the Earth like this, it has concrete implications not just for this generation, but for future generations."

The president made his second post-spill visit to the Gulf the following day, and ordered more cleanup workers to be assigned to coastal areas tainted or threatened by oil. While the president was touring the beaches, a team of federal investigators and prosecutors was assembling in New Orleans to begin the long, deliberate process of gathering evidence for a legal case against BP and others involved in the disaster. And the Coast Guard, the lead federal agency in the response to the spill, announced that it would double its manpower in the Gulf to step up the pressure on BP.

"There is really no excuse for not having constant activity," said Rear Admiral Landry, the on-scene commander, noting that during an aerial tour of the spill she noticed boats and cleanup equipment sitting idle just beyond the edges of the oil slick. "Our frustration with BP is there should be no delays at all" in using every available resource, Landry said. "There is a sense of urgency, and there has been since day one. We have not backed off on this since day one."

It wasn't just idle equipment that was worrying people monitoring the cleanup efforts. While there was little question there were ample resources available, with more than 22,000 people, 1,100 vessels, and 2.5 million feet of boom deployed, how they were being utilized was raising concerns. Signs of sloppiness and even disrespect for the environment were everywhere: booms that were not properly maintained were allowing streams of oil to flow past; yellow caution tape and oil-soaked mops were left strewn on some beaches; response vessels hired by BP had no communications with cleanup team leaders; and on Queen Bess Island, Louisiana, workers were seen stomping through pelican nesting grounds and tossing bird eggs back and forth like baseballs.[47]

"The present system is not working," Senator Bill Nelson (D-Florida) said at a hearing on the response effort. "I still don't know who's in charge," added Billy Nungesser in testimony before the Senate committee. "Is it BP? Is it the Coast Guard?"

Interior Secretary Salazar said on national television that he was "not completely" confident that BP knew what it was doing. "If we find they're not doing what they're supposed to be doing, we'll push them out of the way appropriately," Salazar said.

BP pushed back, with chief operating officer Doug Suttles responding to Salazar on another network by saying, "The federal government does have ultimate control of this event . . . I think what we're seeing is a huge frustration. We share it. I share the secretary's frustration. I want this thing to stop. We want it stopped. The people who live here want it to stop. I'm doing everything we can. I don't know anything we could be doing that we aren't doing. We're getting lots of help from lots of places. The government is looking at what we do every day; they're providing experts as well.

So I think it's this huge sense of frustration that we haven't been able to stop it yet."[48]

Added CEO Hayward in an interview with Reuters: "I understand perfectly why everyone is angry and frustrated that this leak has not been stopped. I am angry and frustrated. I want this thing stopped as fast as I can, as we can. We want it stopped and we're doing everything we can to stop the damn leak and we're going to continue to do everything we can to stop the damn leak. The reality is that it's a very challenging technological challenge."[49]

Amid the growing tension, BP reported that its cleanup costs were soaring in the Gulf, skyrocketing from $6 million a day to $22 million a day in the month of May alone. By Memorial Day, the company said it had spent more than $760 million, and the bills were rising rapidly. But *Bloomberg News* noted in reporting on BP's costs that the company had averaged $45 million per day in profits in 2009.[50]

Some of the money was going into efforts to persuade the public that BP was doing all it could to end the crisis. Full-page ads taken out by the company in major Sunday newspapers in May were headlined, WE WILL MAKE THIS RIGHT.

On Wednesday, May 26, the company began its long-awaited "top kill" on the well, raising hopes around the globe that the end may be in sight. "Let's all keep our fingers crossed, let's all say our prayers," the Coast Guard's Landry said at a news conference. "Everybody's anxious to see success with this intervention."

"If it actually works, we should know fairly quickly during the day," said BP's Suttles in advance of the attempt. "We intend to start the operation in the early hours of the morning. We want to do it in daylight for a lot of reasons including the safety of those on the job. If it works we should get the job complete on that day, on Wednesday. If it doesn't work, I should tell you we have other options. We continue to parallel every option out there, including to contain the flow and actually other options to stop the flow."

The next day, the sense of despair was palpable as the first effort to plug the well with mud and concrete was unsuccessful. "We understand where we stand today," Suttles said on Thursday. "The well continues to flow." Another attempt would be made that evening, he said.

After three failed tries, BP abandoned the top kill idea on Saturday, May 29, and said it would move on to the next option, cutting the well pipe to get a smooth surface and placing a cap on top that would hopefully fit snugly. "This scares everybody—the fact that we can't make this well stop flowing or the fact that we haven't succeeded so far," Suttles said in a somber statement issued by BP. But he expressed hope that the so-called riser cap would be able to capture most of the oil coming out of the well. "We're confident the job will work, but obviously we cannot guarantee success at this time," he said.

Nungesser in Plaquemines Parish said the news that the top kill had failed brought tears to his eyes. "They are going to destroy south Louisiana. We are dying a slow death here," he said. "We don't have time to wait while they try solutions. Hurricane season starts on Tuesday."[51]

President Obama called the latest news "as enraging as it is heartbreaking." But he tried to assure the public the crisis would eventually be resolved. "We will not relent until this leak is contained, until the waters and shores are cleaned up, and until the people unjustly victimized by this man-made disaster are made whole," he said. The president was just back from the Gulf, where he met in Grand Isle, Louisiana, with three governors, two U.S. senators, and other public officials in another bid to show he was in charge. "We're in this together," he told the Gulf leaders.

Later that weekend, as BP was making plans for yet another approach to plugging the leak, two biologists for the Louisiana Department of Wildlife and Fisheries pulled a dead dolphin into their boat in Barataria Pass, not far from where the public officials and the president had met. "It's the freshest one we've seen," said Mandy Tumlin, who had found the six-foot-long male dolphin with her partner Clint Edds just as a storm was blowing in on the Louisiana coast.[52]

Images of dead or oil-covered animals that were starting to saturate the media were beginning to irk BP. There were numerous reports in early June of news photographers and journalists being stopped by BP security guards when they tried to reach locations where oil had hit the shores.

A photographer for the *New Orleans Times-Picayune* was blocked a hundred yards from the shore at Port Fourchon, Louisiana, when he tried to take pictures of tar balls on public property there. A reporter for *Mother Jones* magazine was told that a BP official had to escort her to Elmer's Island, a state preserve, because it was impacted by "BP's oil." A CBS news crew was told it had to leave a public beach by Coast Guard officials who said they were only following "BP's rules."

Journalists attempting to book flights over the spill area also were told they had to get permission from BP's command center to enter restricted airspace. A Louisiana charter service, Southern Seaplane Inc., sent a letter to Senator David Vitter complaining about the restrictions. "We are not at liberty to fly media, journalists, photographers, or scientists," the company told Vitter. "We strongly feel that the reason for this . . . is that BP wants to control their exposure to the press." Reporters also learned from fishermen who had been hired by BP for the cleanup that no members of the press would be allowed on their boats. "You could tell BP was starting to close their grip, telling the fishermen not to talk to us," Jared Moossy, a Dallas photographer, told *Newsweek* magazine. "They would say that BP had told them not to talk to us or cooperate with us or that they'd get fired."

The attempts to control the press coverage were coupled with a major BP ad campaign that included television spots featuring Tony Hayward apologizing for the spill, but saying, "We've helped organize the largest environmental response in this country's history." At the same time, BP hired Anne Womack-Kolton, who served as a spokeswoman for former vice president Dick Cheney and at the Energy Department under President Bush, to take over media relations during the Gulf crisis.

Some public relations experts said media manipulation and messaging would do little to help BP overcome all the negative coverage from the spill. "You can feel sorry for them for one major accident, but they've had three major accidents in recent years: the big Texas refinery disaster, the big oil spill at Prudhoe Bay, and now this in the Gulf of Mexico," Tom Brennan, who helped with media relations during the *Exxon Valdez* spill, told McClatchy Newspapers.

"There's no PR people in the world that can do anything with that, other than try to keep their heads down," he said.[53]

In the face of the attacks, BP executives were getting increasingly defensive. Managing director Robert Dudley went on a *Fox News Sunday* program and was grilled by host Chris Wallace on the company's poor safety record. "In the last three years, the chief executive of the company, Tony Hayward, has brought in a program top to bottom where we focus on safe and reliable operations, ingrained it in the—in the culture of the company," Dudley said.

"Forgive me, Mr. Dudley. That hasn't worked too well, has it?" Wallace asked.

"We've had this accident in the Gulf which we're taking full responsibility for," Dudley responded. "We're not blaming anyone yet for it. The investigation of this will determine the causes. What's happened out in the Gulf today is something that is an industry issue to understand this fail-safe use of equipment. It's going to have implications for the drilling industry not in the U.S. only, but all around the world. Everyone's going to step back and learn from this. . . ."

Wallace then asked Dudley if BP had "cut any corners" in drilling the Macondo well that blew out. "Chris, I don't believe they did," Dudley said. "The casing designs on that well are similar to designs on rigs and wells used all across the Gulf. The decisions that were made in the final hours of . . . decisions to control the well is something that the investigations are going to have to go through in great detail. That's a combination of BP and the rig contractors that made those decisions. Any statement or conclusion before those investigations are out are very premature."[54]

Hayward, facing increasing speculation in the media that he would soon be ousted as BP's chief executive, also had his back up in talking to the press as the leak continued into its second month. He snapped during a visit to the Gulf on Memorial Day weekend that "nobody wants this thing over more than I do. I'd like my life back." The comment was considered so insensitive to the eleven men who died on the *Deepwater Horizon* that Hayward was forced to post an apology to their families a few days later. "Those words don't represent how I feel about this tragedy," Hayward wrote.

Nevertheless, the comment fueled public animosity toward Hayward that in turn created open hostility in the media toward BP and Great Britain in general. The New York *Daily News* described Hayward as the "most hated and clueless man in America," while Sarah Palin made a reference to "foreign oil companies" operating in U.S. waters, apparently forgetting that her husband Todd had done work for BP for nearly two decades in Alaska.[55]

President Obama piled on by telling Matt Lauer on NBC's *Today* show that he would have fired Hayward if he was in charge of BP. "He wouldn't be working for me after any of those statements," Obama said, referring not only to Hayward saying he wanted his life back, but to his frequent attempts to downplay the impacts of the spill. The president also took a harsh tone toward critics of his response to the crisis, saying he was down in the Gulf before "most of these talking heads were even paying attention." He also dismissed the suggestion that he was merely monitoring events rather than taking action. "I don't sit around just talking to experts because this is a college seminar," Obama said. "We talk to these folks because they potentially have the best answers, so I know whose ass to kick."[56]

Attorney General Eric Holder kept the heat on by declaring in New Orleans that he would "prosecute to the fullest extent of the law" if violations of federal law were found to be related to the spill. The president backed up the statement, saying "we have an obligation to determine what went wrong" after meeting with the co-chairmen of a bipartisan commission he appointed to investigate the spill, former senator Bob Graham (D-Florida) and Republican William Reilly, who ran the EPA during the administration of President George H. W. Bush.

Obama also stressed in another visit to the Gulf that he was unhappy to learn that BP was spending millions of dollars on advertisements and proposing to pay dividends to stockholders in the range of more than $10 billion. "I want BP to be very clear they've got moral and legal obligations here in the Gulf for the damage that has been done," the president said after meetings at the New Orleans airport. "And what I don't want to hear is, when they're spending that kind of money on their shareholders and spending that kind of money on TV advertising, that they're nickel-and-diming fishermen or small businesses here in the Gulf who are having a hard time."

After a lengthy setback when the saw became stuck cutting through the pipe of the gushing well, BP engineers working with underwater robots finally made a clean shear at the top of the wellhead in early June. The next step was to secure a cap in place. BP's Doug Suttles explained that the cap would actually be a dual-purpose device that captured oil and kept the pipe from freezing. "It's a pipe within a pipe, and the outer pipe is there so we can actually pump hot water down the outside, between the walls of the riser and the drilling pipe, and we do that to prevent the formation of the hydrates inside of the drill pipe in which the oil is flowing," he said.

Hopes were rising again, though company officials reminded the public that "the beast" had resisted all attempts to tame it thus far. White House adviser Carol Browner was cautiously optimistic. "If it's a snug fit, then there could be very, very little oil," she said. "If they're not able to get as snug a fit, then there could be more. We're going to hope for the best and prepare for the worst."

There were signs of success by the weekend of June 6 after the cap was installed. Retired Coast Guard Admiral Thad Allen, who stepped down as commandant in May but was now serving as "national incident commander" in the Gulf, announced that the device captured about 10,000 barrels of fluids on the first day, or more than half the flow if the most recent estimates of the size of the leak were correct. Allen said the Coast Guard and BP would press on until the well was completely sealed and the spilled oil in the Gulf was fully contained. "And this is a war, it's an insidious war," he said, "because it's attacking, you know, four states one at a time, and it comes from different directions depending on the weather."

With the situation improving somewhat, President Obama reached out to British Prime Minister David Cameron in a phone call on Saturday, June 12, the day the United States was playing Great Britain in a World Cup soccer match. Following their thirty-minute conversation, Cameron's office released a statement assuring the people of both nations that the bonds between them were still strong. "President Obama said to the Prime Minister that his unequivocal view was that BP was a multinational global company and that frustrations about the oil spill had nothing to do with national identity," the statement said.

Other informed observers of U.S.-U.K. relations weren't so kind to Obama. *The Times* of London ran a cartoon under the title "USA v. England" showing Obama wearing a soccer jersey with "Sponsored by Midterm Elections Inc." on the back, kicking a soccer ball labeled "BP."

The same issue of *The Times* carried a commentary by Malcolm Rifkind, a former British Cabinet minister. "The president should make it clear that he has no desire to destroy a great global company," Rifkind wrote. "The future of its chief executive is for the company to decide and not for the White House. While he might, as he has said, wish to 'kick ass,' he should concentrate his energies on more productive activity."[57]

John Napier, chairman of the British insurance company RSA, was more blunt in an open letter to Obama published in the *Times*. "Your comments towards BP and its CEO . . . are coming across as somewhat prejudicial and personal," Napier said. "There is a sense here that these attacks are being made because BP is British."

Obama did not back off the following week in an address to the nation in the midst of four days of White House meetings with top BP executives, including Hayward and BP's chairman, Carl-Henric Svanberg. In a speech from the Oval Office, the president accused BP of "recklessness" and warned that the company faced billions of dollars in cleanup costs. "We will make BP pay," he said.

The next day, Obama announced that BP had agreed to establish a $20 billion fund to compensate victims of the oil spill. The money would be administered by Kenneth Feinberg, a lawyer and mediator

with extensive experience in complex compensation issues. Feinberg had overseen distribution of $7 billion for victims of the Sep-
tember 11, 2001, terrorist attacks; managed a fund for victims of
mass shootings on the Virginia Tech University campus in 2007;
and had been appointed to monitor executive compensation in
companies that received federal bailout funds after the 2008 financial crisis.

The BP fund was the product of hours of difficult negotiations
that ultimately hinged on Obama's ability to convince BP's leaders
that many people were suffering as a result of the company's actions.
"I emphasized to the chairman that when he's talking to shareholders,
when he is in meetings in his boardroom, to keep in mind those
individuals—that they are desperate, that some of them, if they don't
get relief quickly, may lose businesses that have been in their families
for two or three generations," Obama said after meeting with Svanberg. "And the chairman assured me that he would keep them in
mind."

Outside the White House after the meeting, Svanberg seemed to
stumble when asked how it went. The president, he said, seemed
"frustrated because he cares about the small people." Then Svanberg
continued, "People say that large oil companies don't care about the
small people. But we care. We care about the small people."

Once again, a BP executive came off sounding indifferent toward victims of its accidents, if not a little arrogant. BP's public
affairs team tried to argue that Svanberg's remark was a "slip in
translation" by a Swedish native in a difficult situation. Svanberg
himself issued a statement later that day saying he was "very sorry"
that he had spoken so "clumsily." But another firestorm of bitter

commentary directed against BP was already blaring in the blogo-sphere.

Things didn't improve much the rest of the week as Hayward remained in Washington to be grilled by members of Congress about the Gulf spill and BP's role in it. But before Hayward spoke a word as he was seated at the witness table before the House Energy and Commerce Committee, there was a stunning diversion when the top Republican on the panel, Representative Joe Barton of Texas, apologized to Hayward for what he perceived to be a "shakedown" by the White House in pressuring the company to put $20 billion into a fund for victims.

"I am ashamed of what happened at the White House yester-day," Barton said. "I think it is a tragedy of the first proportion that a private corporation can be subjected to what I would consider a shakedown, in this case a $20 billion shakedown."

The veteran lawmaker, first elected to the House in 1984, said it was wrong for the administration to demand a compensation fund at the same time it was weighing criminal prosecution against the company. "There is no question that BP is liable for the damages, but we have a due process system," he said. "I'm not speaking for anyone else, but I apologize," Barton told Hayward.

White House Press Secretary Robert Gibbs responded immedi-ately in a statement. "What is shameful is that Joe Barton seems to have more concern for big corporations that caused this disaster than the fishermen, small business owners, and communities whose lives have been devastated by the destruction," Gibbs said. "Con-gressman Barton may think that a fund to compensate these Amer-icans is a 'tragedy,' but most Americans know that the real tragedy is what the men and women of the Gulf Coast are going through right now."

Later in the day, after being threatened with removal from his Energy and Commerce Committee post by Republican leaders in the House, Barton issued a statement apologizing for his apology. "I apologize for using the term 'shakedown' with regard to yesterday's actions at the White House in my opening statement this morning, and I retract my apology to BP," he said. "I regret the impact that my statement this morning implied that BP should not pay for the consequences of their decisions and actions in this incident."

In between the theatrics, committee members armed with internal documents they had obtained from BP hammered away at Hayward, accusing the company of blatantly ignoring the risks as it rushed to pursue oil in the deep waters of the Gulf.

"There is not a single e-mail or document that shows you paid even the slightest attention to the dangers at this well," charged Committee Chairman Henry Waxman (D-California). "BP's corporate complacency is astonishing."

A senior Republican on the committee, Representative Michael Burgess of Texas, characterized the problem as BP's "general lack of curiosity" about industry safety standards. When Hayward protested that "I wasn't part of the decision-making process on this well," Burgess exclaimed, "But you're the CEO!"

"We drill hundreds of wells," Hayward said. "That's what's scaring me right now," Burgess responded.

Throughout the daylong hearing, Hayward repeatedly told panel members that he could not remember specific events or was not involved in decisions that were being referenced. At one point, Republican Representative Cliff Stearns of Florida wryly asked Hayward if he could confirm that it was Thursday. Hayward said it was.

The day after the hearing, BP announced from London that Hayward was being relieved of his duties as leader of the company's spill response. Managing Director Robert Dudley, a native New Yorker who grew up in Mississippi, was named as BP's new "point man" in the Gulf crisis. BP stressed that Hayward was still the CEO, but was just being freed up to spend more time on the rest of the company's global operations.

BP Chairman Svanberg acknowledged in an interview with Sky News that Hayward had talked himself into an untenable position in the United States. "It is clear Tony has made remarks that have upset people," Svanberg said. "He will be more home and be there and be here, but I think it has been a difficult period and as long as we don't close the well and take care of this, there will be criticisms about many things. Right now that is our focus to make that happen."

Hayward returned to London after his grueling week in Washington and went immediately to the coast to participate in a yacht race with his son. When the story hit the press on Saturday, June 19, BP officials once again struggled to defend their boss's apparent aloofness from the tragedy unfolding in the Gulf. "He is having some rare private time with his son," BP spokeswoman Sheila Williams told *The New York Times*. In a follow-up e-mail, Williams said, "Tony receives regular updates from the Gulf."[58]

Political leaders in Washington were outraged. "To quote Tony

Hayward, he's got his life back," White House Chief of Staff Rahm Emanuel said on ABC's Sunday morning talk show *This Week*.[59]

Republican Senator Richard Shelby of Alabama called Hayward's yachting excursion the "height of arrogance" on Fox News. "I can tell you that yacht ought to be here skimming and cleaning up a lot of the oil. He ought to be down here seeing what is really going on. Not in a cocoon somewhere."[60]

While efforts were continuing to put a tighter cap on the blowout and BP's drilling team kept punching through the sea floor to permanently seal the well from below, the circumstances surrounding the crisis kept getting uglier and uglier.

BP reported in late June that it would begin raising $50 billion to pay some of its spill expenses, through a combination of loans and asset sales. Meanwhile, BP's minority partner on the Macondo well, clearly looking to reduce its own potential liabilities from the accident, issued a statement from its chairman criticizing BP's "reckless decisions and actions" that led to the disaster.

As BP's market value sunk by $88 billion during the spill, pension funds with BP holdings reported massive losses on paper over a period of just two months. For instance, the California Public Employees' Retirement System, with assets of $210 billion, said it lost $284.6 million from its ownership of 58.2 million BP shares, *Bloomberg News* reported. "We will be engaging BP on corporate governance to discuss the impact of the crisis on the value of the company," CalPERS spokesman Brad Pacheco told the news service.[61]

Pension funds in the Gulf region, heavily invested in energy stocks, were ravaged. A retirement fund for Louisiana police officers took a big hit because 20 percent of its assets were BP shares,

The New York Times reported. The state retirement system in Alabama lost nearly $100 million, the newspaper said.[62]

Out in the Gulf, it was reported that sea turtles were being burned alive in controlled fires set by BP and the Coast Guard to reduce oil from the surface. A fishing boat captain told a Florida television station that he witnessed the burnings and was sickened by what he saw. Wildlife groups went to federal court seeking an injunction to ban the controlled burns. In response, BP and the Coast Guard worked out an agreement to let environmentalists scour an area and remove wildlife before future fires were set.

In a particularly nasty diatribe, Sarah Palin tweeted to millions of followers on her Twitter account that they should read an article by conservative writer Thomas Sowell comparing Obama's demand that BP establish a $20 billion compensation fund to Adolf Hitler's pushing the Reichstag to give him dictatorial powers "for the relief of the German people."

"This is about the rule of law versus an unconstitutional power grab," Palin tweeted. "Read Thomas Sowell's article."

Equally provocative was a suggestion for dealing with the blowout made by Soviet physicist Victor Mikhailov in a story by Reuters at the beginning of July. "A nuclear explosion over the leak," Mikhailov said in an interview at the Institute of Strategic Stability in Moscow. "I don't know what BP is waiting for, they are wasting their time. Only about ten kilotons of nuclear explosion capacity and the problem is solved."[63]

By the Fourth of July, tar balls were hitting the beaches in Texas, giving the spill a full sweep across the five states on the Gulf along with Louisiana, Mississippi, Alabama, and Florida. In what could be seen as a touch of dark irony, some of the gooey wads of oil washed up near Texas City, where the explosion at BP's refinery had killed 15 workers and injured 180 others five years earlier.

On Friday, July 9, National Incident Commander Thad Allen announced that BP would be removing the cap on the wellhead, which was now drawing up about 15,000 barrels of oil a day, and replacing it with a tighter seal that might stop the leak entirely. The operation would take a week to ten days and allow more oil to spew into the sea, Allen said, but the trade-off could be an end to the nightmare that had lasted nearly three months.

On Thursday, July 15, a senior vice president at BP, Kent Wells, announced that the new cap had entirely stopped the flow of oil from the well at 2:25 P.M., roughly eighty-six days after the blowout and explosion that ripped apart the *Deepwater Horizon*. Still, Wells tried to contain any celebration. "I am very pleased that there's no oil going into the Gulf of Mexico," he said, "but we just started the test and I don't want to create a false sense of excitement."

The good news, even if only temporary, had a positive effect on BP's financial outlook, pumping up the value of its shares on Wall Street by 7.6 percent in one day to $38.92. And that value was up 45 percent from the stock's fourteen-year low in late June, when a BP share was worth $26.75. "Ending the risk of further spillage is an important step in rehabilitating the stock," Citigroup Global Markets told its clients.[64]

A couple of days later, as the cap was still holding, a Kansas plumber named Joe Caldart told the *Christian Science Monitor* that it was his suggestion for tightening the cap on the well that led to BP's success. Caldart said he initially had no luck getting his idea to

BP soon after the blowout occurred in the spring, and he pretty much gave up. "I was thinking, well, if Kevin Costner and James Cameron were having problems getting through, as famous as they are, I didn't have a chance," he said.[65]

But then Caldart decided to send his sketches for a cap to University of California engineer Robert Bea, who had appeared on *60 Minutes* not long after the accident providing an assessment of what happened on the *Deepwater Horizon*. Bea used his connections with BP to get Caldart's drawings evaluated, and the final design ended up looking much like what he submitted.

"The idea was using the top flange on the blowout preventer as an attachment point and then employing an internal seal against that flange surface," Bea told the *Monitor*. "You can kind of see how a plumber thinks this way. That's how they have to plumb homes for sewage."

"I'm sure we've used bits and pieces of suggestions and have picked things out that could be used going forward," said BP spokesman Mark Salt. Caldart said he wasn't sure he wanted to go public about his role, "but I also felt like people should know that here an average guy submitted something that maybe helped."

While the new cap appeared to be working, it was not the final solution to the crisis. The Coast Guard's Thad Allen insisted that BP needed to complete a "static kill" by pumping heavy mud into the well that would counter pressure from below, and then permanently seal the well from below through a separate relief well that was still being drilled.

BP wasn't finished with being battered by politicians and the media, either. Four U.S. senators from New York and New Jersey revealed in mid-July that they were seeking a State Department investigation into whether BP had pressured the British government to release a convicted terrorist to help the company secure an oil-drilling contract from the Libyan government in 2007.

The senators, all Democrats, said they were concerned about the release by British authorities of Abdel Al-Megrahi, a Libyan who had been convicted for his role in the bombing of a Pan American airliner over Lockerbie, Scotland, in 1988 that killed 270 people, mostly Americans. Al-Megrahi was transferred to Libya around the same time BP received a lucrative contract from the Middle East dictatorship to explore for oil in its territory.

"The question we now have to answer is, was this corporation willing to trade justice in the murder of 270 innocent people for oil profits?" the four senators said in a letter to Secretary of State Hillary Clinton. BP acknowledged that a delay in the prisoner transfer with Libya could have jeopardized its contract, but it firmly denied putting any pressure on the British government to make the transfer, which was being considered because Al-Megrahi was dying of cancer. The British foreign secretary, William Hague, backed up BP's statement, and said there was no evidence the company had improperly intervened. The calls for an investigation were soon dropped, but another message of distrust had been sent around the world about BP.

A frenzy of reports by bloggers in July that BP was doctoring online photos also made the company look bad. The Gawker Web site said it received a tip that one aerial photo of the spill had been touched up to make the sea appear bright blue, rather than oil-tainted, and other photos from inside the company's response center were altered to give the appearance of vigorous activity. BP officials said they were only enhancing, not changing, photographs of its spill activities, but all of the doctored pictures were removed from the Web.

The photo controversy followed news reports earlier in the summer that BP had purchased the rights to a number of Google search terms, including "oil spill," so that the company's own Web sites, with carefully polished explanations of events in the Gulf, would show up in the top results of Google searches for those terms. One California marketing expert, Kevin Ryan, called it a "brilliant move" by the company—a rare accolade in an ocean of criticism about BP's public relations efforts. "The other option BP had was to just not do this and let the news interpret what's going on," Ryan said. "But they're getting so much bad press that directing traffic to their own site is a great PR strategy." [66]

At the end of July, the BP board of directors announced what most observers had been expecting for months: CEO Tony Hayward would be replaced by Robert Dudley, who had taken over from Hayward as BP's point man in the Gulf in June. Hayward would be literally moved to the other side of the globe from the Gulf of Mexico and made a nonexecutive director of BP's joint venture with Russian oligarchs, TNK-BP Ltd., in Moscow. The moves would take effect in October, allowing Hayward time to clean up his affairs in London and giving Dudley time to wrap up his work in the Gulf and learn the ropes at BP headquarters.

"Sometimes you step off the pavement and you get hit by a bus," a visibly shaken Hayward told reporters in London. While expressing his love for BP "and everything it stands for," Hayward said he felt he had been "demonized and vilified" in the United States. "BP can't move on as a company in the U.S. with me as its leader," he said. "I don't know if that will assuage the politicians or not, but it is the right thing to stand down."

In a formal statement, Hayward defended his record at BP and in the Gulf, even as he accepted "deep responsibility" for any failures. "From day one I decided that I would personally lead BP's efforts to stem the leak and contain the damage, a logistical operation unprecedented in scale and cost. We have now capped the oil flow and we are doing everything within our power to clean up the spill and to make restitution to everyone with legitimate claims," he said.

"I believe the decision I have reached with the board to step down is consistent with the responsibility BP has shown throughout these terrible events," Hayward said. "BP will be a changed company as a result of Macondo and it is right that it should embark on its next phase under new leadership."

In early August, the "static kill" on the Macondo well was declared a success, and drilling continued on the ultimate solution, the relief well. Dudley announced that the company would begin pulling back some resources no longer needed in the Gulf, but cleanup of the spilled oil would continue until the job was completed.

Government scientists issued a wide range of estimates of how much oil had been released into the Gulf—anywhere from 94 million to 184 million gallons—though it was later determined that the upper range was probably the most correct. Official reports showed that about 1.8 million gallons of the chemical dispersant Corexit had also been dumped into the Gulf as part of the efforts to reduce the amount of oil that made landfall.

More than six hundred miles of coastline ended up being affected by the spill, and an unprecedented 2,200 beach closings were reported in the Gulf region during the summer of 2010, according to surveys by the Natural Resources Defense Council.

BP reported in September that it had spent $8 billion so far on the spill response, and it reported a massive quarterly loss of $17.2 billion in its second quarter after taking a pretax charge of $32.2 billion to account for the $20 billion compensation fund it agreed to establish.

BP also acknowledged that it had spent more than $1 million a week on TV and radio ads during the crisis, on top of $93 million it spent on a media campaign in the first few weeks after the accident in April.

Hundreds of lawsuits were being consolidated in the federal court in New Orleans after BP's request that they be heard in Houston was rejected. Action on all the cases was likely to take years. A criminal investigation of the spill also continued and federal prosecutors were not expected to decide on charges until well into 2011.

Unrelated to the BP crisis, the Gulf region was rocked on September 2 when an oil platform owned by Mariner Energy Inc. of Houston burst into flames about eighty miles off the coast of Texas. No one was killed and there was no subsequent oil leak, but the accident served as a grim reminder of the dangers still present on the hundreds of drilling platforms and rigs operating in the Gulf.

With the completion of the relief well and a successful sealant job deep below the Gulf, the Macondo well was officially declared dead on September 19, almost five full months after the *Deepwater Horizon* accident. "The well presents no further threat of discharge," said National Incident Commander Thad Allen, in one of his last duties for the government following a forty-year career.

A few weeks later, Billy Nungesser won reelection as president of Plaquemines Parish, and vowed to continue his fight with BP and federal agencies for a full cleanup and restoration in the Gulf. Nungesser and the Lousiana governor, Bobby Jindal, also pressed ahead with plans to build sand berms along the state's coastline, to protect against future spills.

And on October 12, Interior Secretary Ken Salazar announced that the moratorium on new deepwater drilling that had been imposed in April was being lifted. "We are open for business," Salazar declared.

chapter ten

LOST LIVES AND LIVELIHOODS

WHEN BP LEADER TONY HAYWARD dolefully complained that "I want my life back" just six weeks into the response to the Gulf spill, it was as if he had thrust a sword into the hearts of the families and friends of the eleven men who died on the *Deepwater Horizon* on April 20, 2010.

The deaths, presumably by incineration for most, left gaping holes in the lives of scores of survivors, including nearly a dozen young children. One boy would never know his father as a result of the accident—Maxwell Gordon Jones was born twenty-four days after Gordon Jones was killed on the rig.

It would probably be correct to describe the victims as a typical group of Gulf oil workers: rugged and independent, fun-loving and loyal to their coworkers, dedicated to their jobs, and passionate about their families and friends. But of course they were so much more than that, each with unique traits that were seared into the memories of their loved ones, making it impossible for the survivors' lives to ever be the same without their lost husbands, fathers, brothers, sons, or significant others.

Gordon Jones, for instance, was a twenty-eight-year-old Baton Rouge man who had what his obituary described as "a distinctive laugh" that friends and family said they would never forget. A graduate of Louisiana State University in his hometown, Jones was an avid golfer, and always carried a putter with him out to the rig so he could practice his stroke on the carpets during downtime between twelve-hour shifts.

Karl Kleppinger, thirty-eight, was also from Baton Rouge but had settled in Natchez, Mississippi, after serving in the army during Operation Desert Storm and the liberation of Kuwait from Iraqi troops in 1991. During the short war, Kleppinger proposed to his girlfriend, Tracy, by mail and received a response tacked onto the end of her return letter: "By the way, yes." His fellow workers on the rig called him "Big Poppa."

Another Baton Rouge native, Shane Roshto, twenty-two, spent part of every day he was not on the rig with his three-year-old son, Blaine, sharing his love for hunting, fishing, and other activities outdoors, according to Roshto's wife, Natalie.

Shane Roshto had voiced concerns about the BP project before the accident, after the crew had experienced problems for more than a month trying to keep the Macondo well under control. "This is a well from hell," he had told his wife. "He said, 'Mother Nature just doesn't want to be drilled here,'" Natalie Roshto recalled.[1]

Roy Wyatt Kemp, twenty-seven, of Jonesville, Louisiana, was an avid outdoorsman and devout Baptist who had two children, three-year-old Kaylee and three-month-old Maddison, with his wife, Courtney. He had recently been promoted on the rig's drilling team and used the opportunity to arrange a schedule that allowed him maximum time with his two young kids. Kemp was scheduled to be back with them on April 21, and daughter Kaylee had told him she was counting down the days.

Twice already in his life, Kemp had been through brushes with death. At the age of three, he was bitten by a cottonmouth snake. "After that, he wasn't afraid of anything," Courtney Kemp said. Then, in the fall of 2004, he broke his leg playing softball and complications developed that put him in intensive care for a week. Roy Wyatt Kemp was a survivor who finally met his match at sea.[2]

Another father of two—a five-year-old daughter and a one-year-old son—was Jason Anderson, thirty-five, from a small town just west of Bay City, Texas, called Midfield. He had gone into the oil business right out of high school, and was a popular figure in his community—more than a thousand people attended when he married his "soul mate," Shelley Simons, in July 2002. He was working his last shift on the *Deepwater Horizon*, after landing a job with Transocean teaching well control to drillers.

Jason Anderson was a family man to the core, according to his obituary in the local newspaper, the *Matagorda Advocate*: "Jason really did not care too much that the grass was not mowed or the house was not painted. He always said that spending time with his family was more important than whatever chore that did not get done. He always read an extra story, he always snuggled a little longer, and he always stayed an extra day camping if that was what would have made Shelley, Lacy, or Ryver happy. Whatever it took, he would do it. He always danced to every song Shelley asked, whether it was in the middle of Bahula Jungle Road, the garage, El Maton National Hall, or even in their new kitchen. He never, ever ended a phone call without saying I love you. He never got on the plane for work without an extra hug. He never went one day without making someone laugh or smile."[3]

One of eleven children who grew up on a soybean farm not far from the Mississippi River, Donald Clark, forty-eight, tried his hand at the family business for a time but left Newellton, Louisiana, to join the oil industry so he could earn enough to support his four children. "He loved his job," said his wife, Sheila. "The only part he didn't like was leaving." But, she added, "This is a rural area and the only way you could really make a decent living is to leave home." Relatives said Clark was a quiet man, sometimes called "Duck" by his family; his fellow oil workers filled two rows of folding chairs in the Newellton Elementary School gymnasium when a memorial service was held for him on May 8.[4]

Also described as a quiet man, forty-eight-year-old Dewey Revette became uncharacteristically animated in his final hours on the *Deepwater Horizon* when it appeared to him that one of the BP managers was rushing to complete the well. "Dewey got pretty hot," said one of his coworkers who survived the blast. Revette was a thirty-year veteran of offshore drilling for Transocean, but the less experienced manager apparently pulled rank on him. Revette, of State Line, Mississippi, was survived by his wife, Sherri, and two daughters.[5]

Stephen Curtis, forty, was another experienced hand in the drilling room. He grew up in a tiny town called Georgetown in the heart of Louisiana and had the oil industry in his blood. His father, Howard Curtis, was a diver-welder on offshore rigs for thirty-four years; Stephen joined the business right out of high school and had been a driller for seventeen years, taking time away only for a stint in the Marine Corps. Stephen Curtis was a fanatic about hunting and married his wife, Nancy, wearing a camouflage suit; those who attended his memorial service were asked to wear camo in his honor. The couple had two children.

Blair Manuel, fifty-six, was three months away from getting married to Melinda Becnel and was planning a summer honeymoon in New Orleans. He had three daughters by a previous marriage. Manuel had grown up on a rice farm in Eunice, Louisiana, where he was a lineman on the high school football team, playing on both offense and defense. His passion was Louisiana State University sports—he had season tickets to LSU football games and cherished an autographed baseball from the school's 2009 team that won the College World Series.

Adam Weise, twenty-four, of Yorktown, Texas, was also a star lineman on his high school football team and went to work on the *Deepwater Horizon* right after graduating in 2005. He loved to hunt, fish, and drive his pickup truck, making ten-hour trips each way between Texas and Louisiana for the helicopter flights to and from his three-week work shifts on the rig.

Finally, there was Dale Burkeen, thirty-seven, who was a crane operator known as "Bubba" from Neshoba, Mississippi, literally a crossroads village south of Philadelphia. A ten-year veteran of offshore rigs, he and wife, Rhonda, had two children, Aryn, fourteen, and Timothy, six. Burkeen's sister, Janet Woodson, said her brother loved survival shows like *Man vs. Wild* on the Discovery Channel, which the siblings often joked about. "I'd say, 'Bubba, when are you going to be somewhere where you need to survive?'" Woodson said. "And he'd say, 'Anything ever happens to me on that rig, I will make it. I'll float to an island somewhere. Y'all don't give up on me, 'cuz I will make it.' We was hoping that we were going to find him on an island somewhere."[6]

Oddly enough, Burkeen's death was the most mysterious of the eleven fatalities on the rig. He had been operating the crane high above the deck of the floating drilling platform when the first explosion hit and he apparently started to scramble to safety. Witnesses to the series of explosions saw Burkeen thrown from a catwalk about fifty feet to the deck when the second blast occurred, but some said they saw him later trying to help survivors get to the lifeboats, as he had been trained to do. Burkeen, probably injured in the fall, may have sacrificed himself in a last valiant effort to assist others to safety.

All but two of the men who died were employees of Transocean, the owner of the *Deepwater Horizon* that BP was leasing to build its well in the Macondo oil field nearly fifty miles south of Louisiana in the deep waters of the Gulf. The exceptions were Gordon Jones and Blair Manuel, both employed by M-I SWACO, a contractor that specialized in maintaining the mud that was pumped into the well as it was being drilled to keep pressures from oil and gas below at bay.

The job of "mud man" sounds inglamorous, but it is complex work essential to keeping a new well under control before it is completed. The mix of minerals and chemicals in the mud must be designed precisely to withstand a well's high pressures, which can vary according to the types of fluids and underground formations that the drill is going through.

For the most part, work in the drilling rooms was considered intense but safe. "We never worried about Gordon on the rig," said his father, Keith Jones. "We worried about him going out on the helicopter."

Gordon Jones was looking forward to returning home after the end of his shift on April 21, and spoke to his wife Michelle by satellite phone shortly before heading to the drilling room. He reported for work three hours before his scheduled start time at midnight and offered to relieve a colleague around 9:00 P.M. on April 20. Jones probably figured the night would go by quickly, with the well nearing completion and about to be sealed so the *Deepwater Horizon* could move on to its next drilling job and a BP rig could be installed to produce oil and gas from the Macondo well.

Within minutes of starting his shift, though, Jones had to realize there were major problems. There were kicks from the well followed by the unmistakable smell of methane, and the supervisors in the drilling room began making distress calls to senior managers elsewhere on the rig.

Then, it was over, probably instantly for the ten men in the drilling room. A huge bubble of gas and oil burst up the 18,000-foot well, shot through the top, and began showering the rig with fluids and mud. When the gas ignited in a massive explosion, an inferno erupted and it was very likely the men were quickly burned to death in the ball of flames. No traces of their bodies would ever be found.

In discussions later with survivors of the accident, Billy Anderson said he learned that his son, Jason, made heroic efforts to control the pressure surge from the well, allowing time for others to flee. "My boy was cremated," Anderson told *Bloomberg News.* "But the actions he and those other ten heroes took are what made it possible for more than one hundred other people to escape with their lives."[7]

Jason Anderson's job was "tool pusher," offshore lingo for supervisor in the drilling room. He had been part of Transocean's crew on the rig since it went into service in 2001, so he knew it well. "He loved his work and thought of his crewmates as family," said Billy Anderson. In one of the distress calls made before the explosion, a worker told his manager that Anderson was trying to close a cover over the top of the drill to block the flow of gas, or at least divert it out of the drilling room. Apparently the upward force was too great for the cover to stop the leak, but it may have slowed the spread of gas enough to allow many workers to evacuate.

Since none of the eleven bodies was recovered, families and friends of the victims held memorial services in the days and weeks after the accident, with many hoping the tributes would refocus the media's attention away from the oil spill that was beginning to spread through the Gulf.

"I'm following the coverage, but I don't know that I like what I'm seeing," said L. D. Manuel, father of Blair Manuel. "Everyone talks about the birds and the damage to the Gulf and everything, but they never talk about the guys that got hurt. That really bothers me."

"I can see that, definitely, the oil spill is everyone's first priority now," said Nelda Winslette, the grandmother of Adam Weise. "It's such an environmental problem. I understand that. But those eleven men were trying to prevent that spill. Nothing ever gets mentioned about those men, and who they were."[8]

"It seems like people have forgotten," agreed Michelle Jones, even though she had lost her husband Gordon just days earlier. The loss was especially painful for her, not only because it came three days before the Jones' sixth wedding anniversary, but because she was due to have their second child at any time. "He was the glue that bound the family together," Michelle said in what would be one of her last public comments about Gordon. "I've got a lot of good family and support," she added. "It'll be okay someday."[9]

When the memorial service for Jones was held in Baton Rouge, Gordon's father Keith said he noticed a small caravan of black SUVs pull up to the church amid the hundreds of other cars and trucks parked outside. A number of men in dark suits went inside and stayed in the back of the church for the service, then quietly left afterward without speaking to anyone. Keith Jones said later he learned it was a group of BP executives, and included CEO Tony Hayward. "I think he went just so he could say he had attended memorials for the victims," Jones said. "Sure enough, when he testified before Congress, he made a point of saying he had been to the services." But Jones said neither he nor anyone in his family ever heard directly from BP, either by phone or e-mail or written communication.

There was a considerable outpouring of public support for the *Deepwater Horizon* victims, even from people who did not know them. At a memorial service on May 1 for Weise at the Lutheran church in Yorktown, Texas, relatives said it seemed like the entire

community turned out to pay their respects. "It was filled to capacity in the balcony," Weise's sister, Judy Henze, told *AOL News*. "That
was really something, and it showed what a special young man he
was. It was unfortunate he only had twenty-four years."[10]

The public's attention quickly shifted back to the crisis in the
Gulf, though, Henze said. "When I hear anybody talk about it,
they're always talking about the environment," she said. "But this is
also a human tragedy."

Courtney Kemp honored her late husband, Roy Wyatt Kemp,
by arranging to fly out to the site of the explosion about a week
after the accident. She paid tribute by tossing from the plane his
favorite flowers, stargazer lilies, in bunches donated by florists in
their community.

Kemp joined another widow from the disaster, Natalie Roshto,
at a special congressional hearing in Chalmette, Louisiana, in early
June. "My husband and I have two precious daughters: Kaylee,
three, and Maddison, four months," Kemp told the panel of lawmakers. "Our girls will only know what a wonderful father they
had by the stories we tell them.

"While I understand companies must make a profit, I do not
believe it should be at the expense of risking lives and destroying
families," she said. "If proper safety procedures had been taken on
the *Deepwater Horizon*, it is my firm belief that this tragic accident
would have been prevented and my husband and the others would
be alive today."

Roshto, whose husband Shane died on the rig, had filed a lawsuit
seeking damages for emotional distress against BP and Transocean
following the accident, after going into a deep depression from her

loss. "I never thought that I would go home to a bright-eyed three-year-old and have to face the fact that his daddy, my husband, would never come home to us," a somber Roshto told the House committee on June 7. "Every three weeks when Blaine and I would give Shane our last loves sending him off for three weeks, I always feared the helicopter ride, but never did this kind of tragedy come to mind.

"After all the safety schools, meetings, fire drills, and safety regulations I just knew he was safe," she said. "When the events of the *Deepwater Horizon* explosion started to unfold I asked myself, Will I ever personally recover? What if he's out there and they just didn't look long enough? As the days passed, Shane's absence became reality. As the days pass, I ask why? What happened? The life Blaine and I knew is over. My love story came to an end. Though he is a mirror image of his daddy, Blaine now has a void that will never be filled."

A few days later, all of the victims' families were invited to the White House to meet with President Obama. All but one family made it to Washington. A few were skeptical as they prepared for the visit with the president, who had been consumed by the response to the spill and its impacts on the Gulf region. "He's just worried about the birds and the fish and the people who are not making their money," Cindy Shelton, the girlfriend of Adam Weise, told ABC's *Good Morning America* just before the scheduled White House meeting. "And I understand that is their livelihood and they deserve to be taken care of, but what about us?"[11]

Obama met separately with each victim's family gathered in two rooms at the White House. The president "listened for as long as we had anything to tell him," Keith Jones told reporters afterward. He picked up the infant son of Gordon Jones, born three and a half weeks after his death, and commented, "It's been a couple years since I had one this little," Keith Jones reported.[12]

"He stayed until there weren't any more questions for him, or no one had anything else to say," Jones said. "The president's a mighty busy man. And I think we were all pleased that he took whatever time it took."

Keith Jones, a Baton Rouge lawyer, then went back to his primary mission since his son's death: obtaining compensation for his daughter-in-law Michelle and his two young grandchildren for the loss of their husband and father.

Jones and his son Chris, also an attorney, had come across a major legal hurdle while preparing for litigation over Gordon Jones's death. A federal law enacted following the 1912 sinking of the *Titanic* specified that survivors of Americans killed in maritime accidents could sue a vessel's owner for lost wages and funeral expenses for their lost loved ones, but there were no provisions allowing for punitive or other damages. In contrast, U.S. laws passed decades later to spell out the rights of victims in airline accidents at sea made clear that the full range of possible damage claims could be filed against aircraft owners.

Keith Jones believed the exemption from punitive damages in shipping accidents meant that maritime companies could afford to be less rigorous about safety than airlines and other transportation industries. And he knew that in the cases of the eleven deaths on the *Deepwater Horizon*—where no bodies were found so there would be no funeral expenses—BP and the rig owner, Transocean, would be very content to simply pay survivors the lost wages of the men who were killed.

Keith and Chris Jones made it their cause to seek amendments to the Death on the High Seas Act of 1920, and went to Washington four different times in the months after the death of their son and brother to lobby for changes on Capitol Hill. Lawmakers angry about the BP spill didn't require much prodding. The House moved swiftly to approve amendments to the law on July 1, and the bill moved directly to the Senate where it awaited action on an agenda gridlocked by partisan politics. Opposition to the bill was being mustered by shipping companies, but proponents were hopeful they could get the legislation passed once the midterm elections were over in November.

When the bill was being debated in the House, Representative Charlie Melancon (D-Louisiana) displayed photos of Gordon Jones and argued that the nation's outdated maritime laws "encourage companies to take risks, gambling with the lives of workers." Melancon also tapped into public anger about BP by announcing an online petition drive calling for BP to dismiss its CEO, Tony Hayward.

Keith Jones signed it with a bitter comment directed at the embattled BP executive: "My son died aboard the Transocean *Deepwater Horizon*. That's whose life Tony Hayward ought to want back."[13]

Testifying at one of the congressional hearings, Keith Jones acknowledged that getting justice for his son's family was not his only goal. "No amount of money can ever compensate us for Gordon's death. We know that," he said. "But this is the only means available to begin to make things right. . . . Payment of punitive damages by irresponsible wrongdoers is the only way to make them learn. These businesses are there to make money. Punishing them by making them pay some of that money to victims who suffer most is the only way to get their attention."

Later, Jones reiterated that no sum of money could make up for the loss of his son. "Somewhere between lost income and all the money in the world—there's really no fair payment," he said at an event in Washington in October. "But I think Michelle has enough to worry about without ever having to worry about money. When it comes time to settle, that's the approach I'll take."

THERE WERE OTHER VICTIMS FROM the *Deepwater Horizon* accident besides the eleven men who were killed. Among the 115 survivors, dozens later filed suit against BP and Transocean saying that they were suffering from post-traumatic stress disorder, the psychological and emotional distress that often haunts people who have been through war.

One of them was Paula Walker, fifty-six, who had been a laundry worker on the rig and managed to escape with a few bruises. But Walker was still traumatized months after the accident.

Sitting in church in Myrtle Grove, Louisiana, weeks later, a sudden loud noise from the speaker system sent Walker into a panic. "I huddled up on the pew and wouldn't put my feet down because the floor was blue and I thought it was the water," she said. Members of the congregation rushed to her aid and reminded Walker she was not on the rig, but it wasn't the first time a loud sound sent her mind back there. "When you go to sleep, you wake up crying. You wake up with nightmares, thinking the building is exploding," she told the Associated Press. "It's hard coping with the situation I'm in, not going to work, thinking about the guys that lost their lives. It was like a family out there."

A radio operator on the rig, Carl Taylor, sixty-two, said he had similar symptoms. "It was so terrifying. You didn't know if you were going to live or die," he said. "We still have to go on and live with what happened on that rig. It was devastating. I'm sure all of us are still having the memories of that night."

Terry Sellers, sixty-one, said his experience on the *Deepwater Horizon* caused him more anxieties than his wartime action in Vietnam. "The explosion was so tremendous," the Louisiana native said. "I'm still doing counseling about it. I can't handle loud noises right now. . . . If I can ever quit thinking about it, that's when I can move on."[14]

For many survivors of the *Deepwater Horizon*, the explosion was the realization of fears they had harbored for months. Three months after the accident, *The New York Times* obtained reports of worker surveys that Transocean had commissioned before April 20 and found widespread concerns about disregard for safety on the rig. One worker expressed frustration that the mobile platform had never been in dry dock for repairs in all of its nine years of drilling wells on the high seas. "We can only work around so much," the worker said. "Run it, break it, fix it," another said. "That's how they work."[15]

The comments were hauntingly similar to the concerns voiced about safety issues by workers at BP's refinery in Texas City months before an explosion at the plant killed 15 employees and injured 180 others in March 2005.

Top managers on the *Deepwater Horizon* also had their lives shattered by the accident. Federal investigators named at least seven of them as "parties of interest" along with nine companies that worked on the Macondo well, making them all possible targets of criminal charges later.

One was Donald Vidrine, BP's most senior manager on the *Deepwater Horizon*. Twice when he was called to testify before a federal commission investigating the accident, Vidrine's lawyer said he was too ill to attend. Vidrine's friends and neighbors told *Bloomberg News* that he appeared to be agonizing over his decisions on the rig and whether they could have caused the disaster. "They'll look for someone down the line to blame it on, even if that isn't the right thing to do," said Bruce Poret, a Chevron engineer who told a reporter for the news service that he was a close friend of Vidrine. "Don's a great guy and a good neighbor. It hurts us all to see what he's going through."[16]

Other witnesses at the commission hearings testified that Vidrine had argued with one of Transocean's top managers, Jimmy Harrell, about how the well should be completed in the hours before the rig exploded. It was after their heated discussion that Harrell was overheard grumbling that at least there was a blowout preventer on the well to cut off the flow if gas and oil burst through. Harrell was also named a party of interest in the investigation, as was the *Deepwater Horizon*'s commanding officer, Captain Curt Kuchta.[17]

Kuchta seemed to be blamed for an unclear command structure on the rig and for failing to assert his authority and take charge after the explosion occurred. A senior Transocean manager, Daun Winslow, not named as a party of interest, told the Coast Guard and Minerals Management Service panel investigating the accident that Kuchta had to be prodded to order the crew to deploy the rig's lifeboats. Winslow also testified that Kuchta hesitated before disconnecting the rig from the blown well. The delay may have been a reason that the emergency disconnect system failed, possibly causing the rig to sink later.

Another Transocean manager, Yancy Keplinger, also not a party of interest, described a scene of chaos as the rig was being evacuated and laid some of the blame on the captain. Keplinger testified that Kuchta told the crew to lower a lifeboat before Keplinger and Kuchta had a chance to board, forcing both men to jump more than eighty feet into the Gulf. "I was kind of disgusted," Keplinger told the commission, struggling to keep his composure. "I wanted to get in the life raft. I had no choice but to jump."

After a break, Keplinger went on to say that Kuchta offered no help to crew members who were uncertain what to do after the rig erupted in flames. Keplinger said he heard Kuchta tell a group of panicked workers: "I don't know about you, but I'm going to jump."

"I would have expected more from a person of that caliber, that position," Keplinger testified.

Kuchta, knowing he could be facing charges, went into isolation in the months after the accident and declined all requests for interviews. As of the end of 2010, no criminal charges had yet been brought against any party in connection with the accident.

SEAFOOD IS THE SECOND-BIGGEST REVENUE producer in the Gulf of Mexico next to oil and gas; a few statistics give some sense of the fishing industry's broad economic reach in the region:

• There are five times more jobs related to fishing in Gulf Coast counties—777,000—than there are employees in the energy industry in those areas—154,000.[18]

• The dockside value of shrimp, crabs, oysters, and fish landed from the Gulf totals more than $660 million a year, and annual sales of boats and other equipment related to commercial fishing in the region total more than $10 billion.[19]

• About 70 percent of the oysters consumed in the United States come from the Gulf, and nearly half the shrimp.[20]

• Recreational fishermen spend more than $2 billion a year on angling in the Gulf, and generate more than $40 billion in related economic activity.[21]

The directors of six state agencies in Louisiana tossed out some more numbers in a June 2 letter to BP asking for $300 million in immediate aid to the fishing industry. "Preliminary indications based on license sales and data from the Louisiana Department of Wildlife and Fisheries (LDWF) indicate that approximately 6,127 commercial fisherman, 4,238 vessel owners, 645 wholesale/retail dealers, 420 charter captains, 107 marinas, and 1,200 oyster lease-holders managing 358,740 acres of leased water bottoms and 1,047,074 acres of state-managed public seed grounds will be directly impacted," the state officials told BP in describing the effects six weeks after the spill began.

The Louisiana letter said the spill was "causing extreme stress to our citizens in these coastal communities that already have endured five years of recovery from four catastrophic storms." Unemployment and demand for public services were both increasing steadily and were expected to continue rising, the officials said.

But numbers alone don't tell the full story of the BP spill's impacts and how dependent millions of people are on a healthy Gulf ecosystem. The fishing industry's importance to the region took on a heartbreaking human face after the oil slick forced the closure of more than a third of the Gulf's waters in the summer of 2010.

Eric Tiser, a forty-seven-year-old shrimper based in Venice at the southern tip of the Louisiana coast, was featured in numerous TV and magazine reports on the spill because he was one of the few commercial fishermen still trying to making a living trawling in waters outside the areas where the oil had spread. His added expenses for gas and reduced take from the seas were covered to some extent by checks from BP totaling $11,000 during the three months after the spill began. Tiser would have much preferred working for BP on cleanup duty that earned as much as $2,000 a day, but his boat was not one of the two thousand or so selected in the company's Vessels of Opportunity program. And Tiser was not reluctant to tell reporters how he felt about it. "BP has threatened our way of life," he told a blogger for the Natural Resources Defense Council, Rocky Kistner. "But the least they could do is hire fishermen around here like me to clean it up," Tiser said. "Instead a lot of outsiders are getting the work."[22]

When the spill began Tiser was just bouncing back from Hurricane Katrina, which wiped out his house and his four boats in 2005. By late summer he was getting desperate. "I'm broke, I'm going under, and all I'm catching is a heart attack," he told the *Los*

Angeles Times as he stepped off his boat in August. "I've been trying to get on with BP since day one. But they won't take me."[23]

Tiser did get one invitation that month, but it was one that would put him more in the hole. "*National Geographic* magazine is flying me out to a big Hollywood party in celebration of their coverage of the oil spill," he told the *Times*. "I have to borrow $500 from relatives to buy some decent clothes."

Just as desperate was shrimper J. J. Creppel of Buras, up the road from Venice on the Mississippi River delta. Before the spill, he could haul in enough shrimp on a good night to make thousands of dollars. After the leak began to spread toward his boat in Plaquemines Parish, Creppel holed up in his trailer home with his fiancée, watching reports on TV. The underwater video of the thick crude streaming from BP's well literally made him sick—he told NRDC's Kistner he had a heart attack from the stress of watching and spent a week in the hospital.

Later in the summer, with a broken-down boat, no money coming in, and an application for help from BP in bureaucratic limbo, Creppel almost had his truck repossessed, but Catholic Charities stepped in and paid off his loan. There was no money for gas to get to the food banks farther north, though, leaving Creppel to pretty much fend for himself. A fisherman for more than forty years, since he was sixteen years old, J. J. Creppel was thinking about starting a new venture on land that was completely foreign to him—he planned to open a chicken farm.[24]

Michael Roberts of Lafitte, Louisiana, had to hide the tears in his eyes from his young grandson when he headed in May toward his usual fishing grounds in Jefferson Parish, due south of New Orleans. He later posted a detailed description on the Louisiana Bayoukeepers Web site, providing stark contrast to the limited and lifeless reports on the spill coming from BP and the government response teams.

"As we neared Barataria Bay, the smell of crude oil in the air was getting thicker and thicker," Roberts wrote. "An event that always brought joy to me all of my life, the approach of the fishing grounds, was slowly turning into a nightmare. As we entered Grand Lake, the name we fishermen call Barataria Bay, I started to see a weird, glassy look to the water and soon it became evident to me, there was oil sheen as far as I could see. Soon, we were running past patches of red oil floating on top of the water."

He came across another boat hauling booms into the bay and the fisherman described what he had just seen farther out in the Gulf of Mexico: "It was unbelievable, the oil runs for miles and miles and was headed for shore and into our fishing grounds," the man said.

"None of this will be the same, for decades to come," Roberts wrote. "The damage is going to be immense and I do not think our lives here in south Louisiana will ever be the same." He grounded his boat near a bayou and examined the situation on the beach. "The scene was one of horror to me," he said. "There was thick red oil on the entire stretch of beach, with oil continuing to wash ashore. The water looked to be infused with red oil, with billions of what appeared to be red pebbles of oil washing up on the beach with every wave. The red oil pebbles, at the high tide mark on the beach, were melting into pools of red goo in the hot Louisiana sun. The damage was overwhelming. There was nobody there to clean it up.

It would take an army to do it. Like so much of coastal Louisiana, it was accessible only by boat. Will it ever be cleaned up? I don't know. Tears again. We soon left that beach and started to head home. . . .

"My heart never felt so heavy, as on that ride in," Roberts concluded. "I thought to myself, this is the most I've cried since I was a baby. In fact I am sure it was. This will be a summer of tears for a lot of us in south Louisiana."[25]

The waters of Barataria Bay soon became the "epicenter" of the disaster in the Gulf, according to media reports in June. In some ways the tag was fitting, because the vast bay that splits the Louisiana coast into two peninsulas has historical as well as ecological significance.

The legendary pirate Jean Lafitte based a smuggling operation on an island in the bay in the early 1800s, but lost much of his fleet when the British invaded Barataria in 1814. Lafitte returned to his original base in New Orleans and helped Andrew Jackson defend the city from the British the following year, which changed his reputation from "privateer" to patriot.

The bay has declined significantly in the past century as construction of dams and flood-control systems to the north allowed salt water from the Gulf to encroach, destroying some five hundred acres of freshwater wetlands and cypress forests. But Barataria remains one of the richest fishing grounds in the Gulf, teeming with oysters and shrimp, trout and redfish. It is also home to bald eagles, alligators, and some black bears, as well as a primary nesting grounds for millions of migratory birds.

But in June of 2010 Barataria Bay was filling with oil, and the tragedy fueled the anger toward BP felt by many Louisianians, especially the fishermen.

"This is some of the best fishing in the whole region, and the oil's coming in just wave after wave," fishing guide Dave Marino told the Associated Press. "It's hard to stomach, it really is."[26]

Farther east along the coast, another famous fishing community—Bayou La Batre, Alabama—was being devastated by the spill, both from oil contaminating the bayou and from the precipitous drop in business for the town's seafood processors, boat shops, and other fishing-related companies. Featured in the novel and the movie *Forrest Gump* as the location where Forrest helps his army buddy "Lieutenant Dan" set up the Bubba Gump Shrimp Company, Bayou La Batre calls itself the "Seafood Capital of Alabama" because it packages the take for so many Gulf fishermen.

The owner of several processing plants in the coastal community, Walton Kraven, Sr., nearly broke down in July when he told Kenneth Feinberg, the administrator of BP's $20 billion compensation fund for victims of the spill, that the disaster could not have come at a worse time for the seasonal fishing industry. "This is our Christmas," Kraven said.[27]

More than two hundred residents of Bayou La Batre jammed into a meeting with Feinberg at 7:00 A.M. on a Saturday in late July and many vented their anger about the spill. At the end of the hour-long session, Feinberg said, "I learned today the depth of frustration in people here on the coast." He went on to try to reassure the residents that he would do all he could to help the community stay afloat. "I am your lawyer," he said. "I do not work for BP. I do not work for the White House. I work for and answer to the residents of the Gulf."

"Feinberg is full of baloney," a bitter Delane Seaman told an Associated Press reporter after the meeting. "He is a lawyer and that is how lawyers talk. I do not believe a word he says. . . . BP is telling us we will be compensated for 100 percent loss of our oyster processing business, too. It will not happen."[28]

Back in Louisiana, state health department officials were being inundated with people who needed counseling on how to survive without spiraling into destructive behavior.

"Most people are in disbelief," said Dr. Tony Speier, deputy assistant secretary of the state's office of mental health. "There's fear not just for economic survival, but for a way of life."[29]

The Louisiana health experts were well aware that there was a surge of substance abuse, domestic violence, divorce, and even suicide among residents of Alaskan coastal communities for years after the *Exxon Valdez* spill in 1989. Counselors for the state agency, along with hundreds of others working for local clinics and nonprofit groups, tried to reach out to fishermen and their spouses and offer free consultations on coping with stress.

One woman who agreed to counseling, Rachel Morris, told *The New York Times* in June that her husband, a fisherman who signed on with BP for the cleanup, went into a depression and took to drinking excessively after seeing the damages on the Gulf. The man, thirty-four-year-old Louis Lund, Jr., told the reporter that the spill and its effects were haunting him, even in his New Orleans home well upstream from the Gulf. "If you're not out there in it, you can't comprehend what this is about," Lund said. "We're going to be surrounded by it. You're going to smell it right here."[30]

More than a third of the fishing community in the Gulf—and by some estimates as many as half of all fishermen in the region—came to the trade from Vietnam after the war there ended in 1975. Many of the "boat people" who fled the country after the fall of Saigon were welcomed to Gulf communities by American assistance groups, particularly Catholic Charities. A former archbishop in the New Orleans diocese, Philip Hannan, reached out to Vietnamese refugees scattered around the country and invited them to move to the city. Thousands poured into the region during the 1970s and 1980s, and by 2010 more than 20,000 were working in an industry they knew well from their native country. And fishing enabled many of the new Vietnamese Americans to maintain their own culture, including their language, which meant that many never bothered to learn much English.[31]

The language barrier added to the problems for Vietnamese fishermen after the spill hit, making it harder for them to understand why they were suddenly forced to end their livelihoods and depend on an oil company or relief agencies to provide them enough money to survive. At the Mary Queen of Vietnam Church in New Orleans, the Reverend Vien Nguyen summed up the situation for CNN in June: "These are proud, active people who contribute to their own livelihood, and now they have to be in lines," the pastor said. "It is a devastating blow."

The deputy director of the parish's Community Development Corporation, Tuan Nguyen, explained the problem in more personal terms. "One of my wife's uncles is a very proud man," he said.

"He's a deckhand. I told him to come in and talk about services. He said, 'I can't stand in line. What if someone sees me?'"

More than 1,500 Vietnamese sought counseling in early May for depression, and a Vietnamese American who represented many of them in Congress, Representative Anh Joseph Cao (R-Louisiana), set up a "rapid response team" through his office to provide translators for people working to assist the community.

"I spoke to a group of fishermen, mainly Vietnamese Americans, and a group of them came up to me . . . they told me that they contemplated suicide because they're in such despair," Cao told a New Orleans television reporter in May. "For some people, this is almost a boiling point where they can no longer handle it and they're going to crack."[32]

"I feel like I am lost," Minh Chu, a fifty-two-year-old deckhand on a fishing boat that was docked by the spill, told an Associated Press reporter in July. "Sometimes I worry and I cannot sleep. I'm thinking about how am I going to make money to sponsor my wife, thinking about how am I going to pay my bills."[33]

Chu had come to New Orleans in the 1970s and worked in a factory before landing his first fishing job. He married a Vietnamese woman in 2007 and had been sending her money since then to save for a new life in America, but after the spill he barely had enough money to buy food. His main source of support was an occasional voucher for groceries from one of the local charities, which required lining up with dozens of other out-of-work fishermen early in the morning, waiting for a handout. For some, the experience revived feelings of hopelessness and despair they still harbored from the war that tore apart their homeland for two decades.

"They've experienced turmoil before and come through it, so there's a certain confidence they will prevail," said the Mary Queen of Vietnam pastor, Vien Nguyen, in July. "On the other hand, some people are overwhelmed and I don't know if they have the time and strength to rebuild. I'm still waiting to see how it turns out."

Caught in a no-man's-land from the oil disaster was the Pointe-au-Chien tribe that lives in the rich bayou west of Barataria Bay. The nearly seven hundred French-speaking Indians are not recognized as a tribal nation under federal law, stripping them of many rights that most other Native Americans enjoy, including the power to defend their land against incursions by oil companies.

A New Orleans attorney who has done work for the tribe on aboriginal land claims, Joel Waltzer, said the Pointe-au-Chien were left out of the rush to stake land claims after the Louisiana Purchase in 1803. "They were not English-speaking, they were completely illiterate, and they had no means to make it to New Orleans and make their claim," Waltzer said.

Even tribes with federal recognition, such as the United Houma Nation, had seen their southern Louisiana settlements decimated by the growth of the energy industry over the past century. "This is not a two-week story, but a hundred-year story," Houma historian Michael Dardar told the Associated Press at the height of the spill. "Coastal erosion, land loss, and more vulnerability to hurricanes and flooding all trace back to this century of unchecked economic development."[34]

Effects on the lands from oil and gas pumping and leaks from wells had rendered agriculture obsolete for many Louisiana tribes, leaving them almost entirely dependent on fishing for income and sustenance. Then came the BP spill, and life for many Indians on the coast turned desperate. Some found jobs in the cleanup effort, but those who didn't had difficulties getting disaster aid because of language barriers and paperwork problems. A flood of suspect claims filed with BP also made it harder for Indians to get reimbursement, the Pointe-au-Chien's chief, Chuckie Verdin, told the *Epoch Times* in October. "People who were not in the fishing business came in and wanted to make claims, and BP officials saw that and started getting stricter," Verdin said.[35]

One member of the tribe, Russell Dardar, was invited to fly to Alaska in August to meet with fishermen affected by the *Exxon Valdez* spill and get some advice on how to survive the oil crisis in the Gulf. Dardar told an Alaskan television station when he arrived in Anchorage that he was lost. "I'm—I was a fisherman, my daddy was a fisherman, his daddy was a fisherman, and so on and so on," Dardar said. "The erosion—we ain't got farming no more, but all we have left was fishing. And now since the oil happened, we don't have much of that."[36]

In mid-August, shrimping was reopened on the bayou, but the Pointe-au-Chien remained in limbo. "Not many fishermen have returned to fishing," said an August 17 report on the tribe's Web site. "There has been oil spotted below Pointe-au-Chien, and there has been no testing by the EPA as to the whether there are dispersants in the water."

Some of the clearest spill damages came in the oyster business, a $1 billion a year industry in the United States with more than two-thirds based in the Gulf of Mexico. All indications were that 2010 was going to provide a rich harvest, until the Macondo well exploded on April 20.

Within a month of the accident, fragile oyster beds that depend on the right balance of salt water and fresh water were decimated from two directions. Those in coastal waters hit by the slick were destroyed by petroleum. Beds in the Mississippi River basin experienced huge die-offs when state officials opened the floodgates and allowed more fresh water to flow toward the Gulf in an attempt to keep the spill from moving into the wetlands. The strategy worked to some degree, but millions of oysters on the shallow bottomlands lost the vital salinity needed to survive and were wiped out.

Unlike shrimpers and other fishermen, oystermen did not have the option of moving to areas unaffected by the spill to try to fill their boats. "Are you going to take care of all the oysters I lost?" an angry Anthony Zupanovic asked BP officials who held a meeting in Plaquemines Parish in late May to pledge assistance for anyone who suffered financial losses from the spill.

When BP executive Bob Fryar promised that the company would compensate Zupanovic, he was not convinced. "It makes you want to throw up when you see it," Zupanovic told a *New York Times* reporter, after describing how oil was moving into his beds in the Mississippi River. "Because you know it's coming and you can't do anything about it."[37]

Louisiana officials reported in August that in the four months since the spill began, a little more than $8 million worth of oysters were harvested in state waters, an amount nearly two-thirds below the oyster haul in the same period one year earlier.

The losses hit especially hard on the Louisiana Oysters Association, a coalition of mostly black families who had thrived on the
oyster trade for generations. Many were descendants of slaves who
became sharecroppers after the Civil War. Over the next century,
the families eked out a living growing potatoes and okra, and some
gradually saved enough to buy their own boats and enter the fishing
business. "For a period of time, we were doing good," association
president Byron Encalade told *Time* magazine in June. The vicious
hurricane season of 2005 damaged many of the oystermen's homes
and boats in delta towns like Phoenix and Pointe a la Hache, but
the close-knit communities had bounced back by 2010 and were
looking ahead to a banner year.[38]

The spill then turned life around completely for oystermen like
Bernard Picone, who had worked the waters from his boat based in
Pointe a la Hache for two decades. Picone, forty-two when the spill
hit, had always been his own boss, but with the public oyster beds
that he usually trawled either closed or decimated by June, he and
his three-man crew signed up with BP for the cleanup and spent
the rest of the summer on the boat, taking on whatever assignments
they were given in the far reaches of the Gulf.

The pay was good—$1,500 a day—but the contract was ended in
early October after BP capped the well and removed most of the
surface oil from the Gulf waters. Picone's boat was back in the dock,
and he was left to wonder when the oyster beds would be thriving
again. The destruction of millions of oysters meant reproduction
would be slowed, and there was speculation it could be at least
three years before the population rebounded.

Picone told the *New Orleans Times-Picayune* in October that he was troubled that leaders of the Louisiana seafood industry were focused mainly on marketing Gulf products to a wary nation rather than figuring out what needed to be done to restore the fishery itself.

"I've been doing this a long time, and you've got good years and bad years with any seafood. It's just like farming," he said. "But they want to talk about 'promote, promote.' Well, what if we ain't got nothing to promote?"[39]

Oyster processors in New Orleans had exactly the same fear. At the P&J Oyster Company just outside the French Quarter, owner Al Sunseri told the *Guardian* newspaper in September that only about 15 percent of the state's oyster beds were truly productive after the spill, even though nearly half the acreage was officially open.[40]

The crop coming in was of much poorer quality than in most years, he said. "We don't have babies, and we don't have the market-sized ones," he said. Pointing to a table covered with oyster shells about half their normal size, he said, "These would ordinarily not be harvested for another year. They really should be in there developing. The few little oysters that I am selling right now are really inferior."

Another large segment of the fishing industry hit hard by the spill was the charter business, which provides a sizable share of the estimated 23 million excursions into the Gulf by recreational anglers each year. In most years, charter boat operators love the oil and gas industry, with some even working part time on the side shuttling supplies to offshore rigs. The rigs themselves are favorite destinations for fishermen seeking prize catches like dolphin, tuna, and blue marlin, which are attracted to all the food sources that cling to the underwater support structures.

The BP spill dampened business on most charters in the northern

Gulf for months. Ryan Lambert, who had well over $1 million a year in charter bookings before the spill, had mostly cancellations afterward. To make matters worse, BP assessed his losses at only $66,000, forcing him to spend hours with his accountant pulling together paperwork. "I shouldn't have to fight for the money that is owed me," he fumed. "I am not the bad guy here. They are the ones who ruined it for me, not vice versa. For me to have to fight for them to pay me for what they did makes me sick."[41]

Damon McKnight, operator of Super Strike Charters in Venice, was in a similar boat. He received $5,000 from BP in May, while two of his captains received $2,500 and his deckhands each received $1,000. In a good month during the summer fishing season, McKnight said his company took in at least $50,000, providing a decent living for all of his employees.[42]

A number of charter boat owners landed jobs in BP's cleanup program, which some of them considered a mixed blessing. The pay was good, but Johnny Nunez, who lost all his business on Fishing Magician Charters in Shell Beach in May, was depressed by his assignments in the spill zone. "We've seen thousands and thousands of dead fish," he said in October, seated in a rocking chair on the front porch of his home. "There also were dead turtles, even a dead whale."

A fisherman for forty-five years, since the age of ten, Nunez and his three sons shuttled people in the response teams around the Gulf, then did some cleanup work. But Nunez felt the efforts were poorly managed, with Coast Guard officials telling crews not to bother going into grassy areas of marshes where oil had spread because the task was too difficult and time-consuming.

Once fishing was restored in the fall, Nunez stayed in the dock.

"I wouldn't eat them," he said, "not when they're letting them spray chemicals that aren't allowed in other countries." One of the most commonly used oil dispersants, Corexit, was banned in Great Britain, and Nunez said he understood why. "The chemicals burn you," he said. "They tell you, 'Don't get in the water, don't touch the water without gloves,' and now they're saying it's okay to fish? It's a contradiction."

Nunez was worried, too, about his community in the eastern reaches of the Louisiana delta, which was torn apart by animosity over who got BP jobs and who did not. Commercial fishermen felt the contract work should be reserved for them, and not given to charter boat operators; fishermen who were spurned by BP were angry with those who were picked.

"Shell Beach is like a family," Nunez said. "Most are from the Canary Islands—my great-grandparents came in 1789 from there. We're all born and raised together, but it's been getting ugly. It's gonna get worse, too."

There were troubling problems inside many homes as well, said Nunez's wife, Karen, a librarian in the local school system. "The men are the foundations of the home," she said. "With Katrina, the men could still say, 'Okay, I'm gonna go to work.' Now, the men are scared."

The fall shrimping season in the Gulf opened as usual in mid-August, and about 80 percent of the waters were now open to fishing, up from around 63 percent at the height of the spill. The government's early tests on Gulf seafood found no traces of oil, and federal officials wanted to spread the news. "We need to let the American people know that the seafood being harvested from the

Gulf is safe to eat," said Commerce Secretary Gary Locke during an August 16 visit to Louisiana. "I think there have been a lot of misperceptions out there. A lot of testing is done before we open state and federal waters to fishing. We're being very thoughtful, very careful, and very deliberate."[43]

Many fishermen, anxious as they were to return to their trade, were skeptical. "We are only going to get one shot at this," said shrimper Acy Cooper. "If we don't do it right, we are going to be in big trouble if any tainted shrimp gets on the market. We don't want to get anything on the market that is going to kill us in the long run."

"Fishermen here are calling it 'voodoo seafood' because we are all cursed," said Bill Thompson in Long Beach, Mississippi, just outside Gulfport. "We do not think it is safe but the state officials say it is. Who do you trust? The people that know these waters or the government?"

Even if much of the oil that leaked from BP's well had evaporated or been cleaned up as the government maintained, there was still a lot in the water and fishermen knew what they had been seeing all summer. Some noticed that fish and crustaceans were swarming toward the surface in some areas, as if fleeing from oil plumes down below. "It looks like all of the sea life is trying to get out of the water," said fisherman Stan Fournier in Alabama. "In the forty years I have been on these waters I've never seen anything like this before."[44]

Crabbers in Mississippi were reporting black stains under the shells of their catch in August, and shrimpers were reporting similar findings, as were some scientists. "They have seen oil in the gills of shrimp," said William Hogarth, dean of the College of Marine

Science at South Florida University, during a November conference on the Gulf spill. "There may not be an immediate effect on species right now, but we could be seeing such an effect in a year, three years, five years from now."[45]

Even a few federal officials were being cautious about assessing the oil's effects on marine life in the Gulf. "I think it's fair to say we won't know for some time yet the full impact," said Jane Lubchenco, head of the National Oceanic and Atmospheric Administration. "Many of the suspected impacts will be on the juvenile stages, the eggs or the larvae, for example, of fish, but also crabs, shrimp, other species. And it's very difficult to detect as it's happening."[46]

With so many reluctant fishermen, and so many who were occupied with the spill cleanup for most of the summer, the nearly two hundred seafood processors on the Gulf Coast were operating at about 20 percent capacity in October. "It's slowed way down," Tony Lyons, owner of Southern Aire Seafood in Irvington, Alabama, told *USA Today*. "We may not be open by the first of the year."[47]

Many of the public oyster grounds in Louisiana were still closed in October, with vast areas of the shallow reefs filled with dead oysters.

The chairman of the Louisiana Oyster Task Force, Mike Voisin, feared it could take three years before the crop returned to normal. His own company, Motivatit Seafoods in Houma, was processing about half the number of oysters in 2010 as it did the year before, but because prices were hiked significantly on the supplies that were available, the company's revenues were level with 2009, Voisin told *Bloomberg News* in September. "I'm not as much worried about this year," he said. "Next year is where my greatest fear comes."[48]

Voisin said his biggest concern was that when the oyster crop rebounded, consumer acceptance might not. "Prices will drop. We'll be forced to discount, but if the market won't accept [our product], we'll have an even greater catastrophe. So the worst may be yet to come; a balloon in supply, without demand catching up." That's why he felt it was important for marketing to be increased nationwide, with BP picking up the tab, Voisin said. "We need a commitment, and we need a major investment by the people that created the challenge," he said. At that point BP had agreed to provide $13 million over three years to test oysters for any problems, but Voisin felt much more would need to be done to aggressively promote the products.

The head of Louisiana's Seafood Promotion and Marketing Board, Ewell Smith, agreed. "We need support," he told the *Daily Comet* in Thibodaux, southwest of New Orleans. "We need to be able to do the same thing BP has done. They've spent millions rebuilding their own image. This board has spent millions building up the brand of Louisiana seafood, and the BP gusher pulled it down."[49]

Smith noted that Alaska spent $100 million over ten years to shore up its fishing industry after the *Exxon Valdez* spill in 1989, and the public focus on that disaster was far less intense than media coverage of the Gulf spill. "This is a $30 million to $40 million a year problem for at least five years," he said.

Louisiana officials kept up pressure on BP to provide funds for marketing Gulf seafood. The secretary of the state Department of Wildlife and Fisheries, Robert Barham, wrote BP's incoming CEO Robert Dudley in September asking for a five-year commitment of

$173 million to test and market Gulf products. Barham cited polls showing that half of Americans believed there was a risk of getting contaminated food by eating at Louisiana restaurants, and anywhere from a third to a half planned to avoid eating fish or shellfish from the Gulf.

Although it pained her to say so, one of the family owners of Prestige Oysters in Texas, Lisa Halili, admitted she could understand the public's fears. She compared the perception of oysters to the taint attached to Tylenol in the 1980s after a poisoning scare involving the over-the-counter drug. "I have to be honest," Halili told *The Wall Street Journal* in November. "I didn't want to use Tylenol."[50]

THE FISHING INDUSTRY GARNERED THE most attention during the crisis in the Gulf, but there were thousands of other businesses affected to varying degrees, from hotels and restaurants that lost most of their guests to churches that saw their Sunday collections drop in half.

"I've been through Hurricane Camille, Hurricane Frederick, and Hurricane Katrina," real estate agent Greg Miller in Gulf Shores, Alabama, told *Bloomberg News*. "They all pale in comparison to this." Suddenly nearly all the bookings at some one hundred vacation properties Miller managed were dropped within weeks after the spill began. "People are canceling left and right," he said in early May. "The phone is ringing off the hook."[51]

Another realtor in Gulf Shores, Bill Brett, lost bookings on 117 coastal condominiums in one fell swoop. A Texas church camp program called Wild Week decided that the 750 people who had signed up would enjoy their "beach camp" more in Missouri than in a potential disaster area. "With tar on the beach, the churches didn't want to go," Harold Hanusch, a leader of the Wild Week program, told *The Wall Street Journal* in June.[52]

Hundreds of other property managers along the coast facing similar losses joined a class action lawsuit seeking reimbursement from BP. "There are questions of diminution of value, cleanup costs, and all kinds of issues for the population and condo owners," said their attorney, Robert Cunningham of Mobile, Alabama. "The closest thing is the *Exxon Valdez* incident. There's a lot more property threatened here."[53]

Hotels and rental properties in areas where BP set up staging operations for the cleanup were filled with workers for months after the spill, but they were not the typical guests who spent money at beach businesses such as restaurants and tourist shops. At Gulf Islands National Seashore near Gulfport, Mississippi, for example, the concession stand at the usually popular Ship Island was idle at the beginning of May. Owner Louis Skrmetta told the *Los Angeles Times* that he usually sees more than a thousand visitors on a weekend day in warm weather, but even school groups had canceled plans for tours of the national seashore, though there was no oil on the beaches yet. "We are wiped out—this could bankrupt me," Skrmetta said.[54]

The U.S. Travel Association, representing thousands of tourism-related companies nationwide, estimated that vacationers spend about $34 billion along the Gulf Coast every year, supporting 400,000 jobs in the region. A few months into the disaster, the association commissioned a study that forecast losses for that sector of the economy exceeding $23 billion over the next three years. "The oil spill will have long-term effects on businesses and jobs in the Gulf Coast region unless we counteract the usual course of events with an unprecedented response," said Roger Dow, the association's CEO, in July.[55]

The group said that if BP immediately provided $500 million to promote travel to the region, the impact might be reduced by $7.5 billion. BP did end up providing about half that amount to Gulf states and tourist businesses experienced a significant rebound in the fall. But in the summer months, beachfront hotels reported that revenues had dropped as much as 30 percent from the peak season in 2009, and some smaller tourism businesses had even greater losses. Endless news reports about the oil spill "essentially shut down tourism for the peak season of the year," said Violet Peters, head of the Jefferson Parish Convention and Visitors Bureau in Louisiana. "Most of our small businesses are tourism related," Peters told the nonprofit CNSNews.com in October. "In August 2009 there were thirty charters. In August 2010 there was one. It's been devastating."[56]

In the early weeks of the crisis, BP provided emergency checks to help hundreds of small companies, especially in the fishing industry, pay some bills and stay on their feet until business started to turn around. Most fishing boat owners received at least $5,000, and similar amounts went to boat-repair shops and seafood processors who lost much of their work in May. But even though BP had more than five hundred claims adjusters working in twenty-two emergency aid centers across the region, many businesses indirectly affected by the spill had a harder time obtaining help.

Restaurant owners Matt and Regina Shipp in Orange Beach, Alabama, estimated that they lost $10,000 just in the last ten days of April after news of the spill spread and people avoided the coast like a plague. But the Shipps were told by a BP adjuster in mid-May that emergency checks were only being issued to fishermen and that restaurants were not eligible. Matt Shipp went to another adjuster, who rejected the claim because the Shipp's Harbour Grill had been open for less than two years, making it hard to determine if the lost business was really related to the spill. Shipp refused to give up, and finally had his first claim paid in mid-June. "Only because of my extreme persistence (four days of eight to ten calls per day)," was he able to get some aid, he told Alabama officials in a letter of complaint later. By that time the restaurant's losses had mounted to nearly $60,000, and Matt Shipp was beginning the process all over again to seek more reimbursement from BP.[57]

When President Obama announced in June that BP would establish a $20 billion fund to compensate victims of the spill, the attorney appointed to administer the fund, Kenneth Feinberg, pledged to process claims rapidly. "We'll do it better than BP and we'll do it quicker," Feinberg told a crowd in Alabama seeking information about the reimbursement program in July. But as thousands of claims piled up on his desk, it became clear that Feinberg and his staff would have difficulty delivering on that promise.[58]

Some claims broke new ground for defining business losses. The pastor of the Anchor Assembly of God church in Bayou La Batre, Alabama, Dan Brown, asked for $50,000 from BP based on a $12,000 drop in donations over several weeks in June, plus anticipation that the collection plate would be $38,000 lighter over the next year.

Brown said fishermen in the coastal community were struggling to feed their families in the wake of the spill, and could no longer afford to give to the church. "You can't tithe what you don't have," he said. "We're fighting for our lives just like a business."[59]

Others sought help from the fund by arguing that the spill had a domino effect on businesses throughout the Gulf, and that they were left at the end of a cascade of economic losses with nowhere to turn in a recession.

In Destin, Florida, at the center of the Panhandle over the Gulf, Christine Watson filed for assistance because she and her family were about to be evicted from their apartment in August. She and her husband lost all their income earlier in the year when their small construction company went under, and Watson took a job at a health and fitness store. But when tar balls started washing up on the Gulf beaches in June, frightening most visitors away from the coast, the drop in business forced the health store owner to lay off all the employees. "We were very financially stable," Watson told the *Washington Independent* newspaper in September. "Now everything has changed. We have spent everything we have. I'm forty-six and I've never been through anything like this." [60]

A property manager in Gulf Shores, Alabama, Kay Hasting, filed a claim for $18,000, which she estimated were the losses she incurred after all but one booking she had lined up for beach properties in June and July were canceled. Hasting said in September she was three months behind on her rent, facing cutoffs of her phone and Internet service, and sadly, unable to send her son to Tennessee for his grandfather's funeral. "My life is on hold," she said. "I can't do anything."

One of the most unusual claims for compensation came from two sisters in Orange Beach, Alabama, who said their wedding planning business had been ruined by the local economic collapse that followed the spill. Sheila Newman and Sheryl Lindsay said fifteen contracts for organizing weddings in the summer were canceled, and no new bookings were being made. Orange Beach Weddings was forced to sell off assets and vacate its office space, and only a cash payment from BP would keep it alive. "This is our lives," Newman told National Public Radio in August. "We live here. This is our business, and our families are here. Are we supposed to just move to another area and forget about what they've done?" BP ended up reimbursing the sisters about $29,000, but Lindsay estimated their losses were nearly ten times that amount. "I'm scared that BP is going to pull out and leave us hanging with nothing," she said.[61]

Everyone who sought help from the BP compensation fund would face a very difficult decision, even if they were offered money, requiring them to essentially make a gamble on their future. Once they accepted reimbursement from the fund, they would forfeit their right to sue BP later on even if the amount they received did not cover their losses.

Fishermen were in an especially tough dilemma, since none of them knew for certain when conditions in the Gulf would return to normal or how long it would take for consumers to overcome fears about possible contamination from the region's seafood. Hotel and restaurant owners had to make similar calculations about when tourism would rebound, plus they were being told by the fund administrator that he was going to carefully scrutinize claims from the tourism industry, which he considered the most "problematic" ones to judge. "I'm going to have to draw some tough lines," Feinberg said.

Still, Feinberg encouraged anyone who suffered financial losses from the spill to submit a claim and get a sense of how they might fare in court. "Simple solution: File a claim, find out how much I'll give you, and then decide. . . . It's almost like you get a free preview," he said. "If the claimant ultimately feels that he or she can do better by filing a lawsuit and litigating for years in federal court, that is the claimant's option. And we will see."

Florida Attorney General Bill McCollum raised concerns about the process in August, telling Feinberg in a letter that it "appears to have as its primary goal the reduction or elimination of claims against BP, instead of making claimants whole."

A lawyer for restaurants and hotels in New Orleans, Stephen Herman, called Feinberg "a glorified claims adjuster for BP," adding that "there's a tremendous risk that he's going to settle this thing as cheaply and as early as possible, and five years down the road, when you've still got environmental damages and all the settlement money is used up, you could have a bunch of people who've been taken advantage of and have nowhere to turn."[62]

Some restaurant owners didn't bother seeking compensation and went directly to court against BP. Chef Susan Spicer, owner of the renowned restaurant Bayona in the New Orleans French Quarter, filed suit in June on behalf of a group of restaurants that claimed they were harmed by the spill. "I have great confidence in my local vendors and the local products that I am serving," Spicer said. "But

I know my suppliers are suffering from the reality of a diminished supply and the perception that all Gulf seafood is unsafe." BP, she said, "needs to be held accountable for its negligence."[63]

Some restaurants miles away from the coast filed similar litigation. The Fish Market Restaurant chain based in Birmingham, Alabama, went to federal court in August claiming losses related to the Gulf spill. Owner George Sarris said reduced supplies and higher prices of seafood added more than $200,000 to his business costs, and he did not believe he would qualify for the compensation fund because his company was not located on the coast.[64]

While the bulk of litigation against BP related to financial harm from the spill, a number of lawsuits were filed claiming health damages from exposure to oil or chemical dispersants—or both—particularly during the cleanup effort.

Joe Overstreet of Fairhope, Alabama, said his work skimming oil from the Gulf left him with rashes, body aches, and frequent headaches. "I take Benadryl pretty much every night so I don't wake up with a headache," he said. "I have pains on my right side recently, and unbelievable headaches. When they start happening I have to stop everything. I have them every day."[65]

Clayton Matherne, a thirty-five-year-old native of the Louisiana bayou, became ill almost as soon as he started working with a cleanup crew on a boat in May, and doctors and lawyers told him later he showed signs of exposure to benzene, a toxic component of crude oil. Matherne was laid up for weeks afterward, constantly taking medication to help him breathe.

"You get a good whiff of some of these chemicals, and it can cause acute immediate symptoms, and more often than not it usually goes away," said Houma, Louisiana, attorney Duke Williams, who specializes in environmental cases involving the oil industry. "But no one fully knows what we are dealing with out there. On the back deck of a vessel it's hot, you're dragging boom, condensing this stuff, the wind is blowing up over the vessel, and human beings react to things differently. They are acutely affected, they go to the emergency room, and are sent home, and these are pretty tough people. They think it's a cold or the flu, and they fall through the cracks."[66]

At an October rally of fishermen in Baton Rouge, several said they were still experiencing health problems from their exposure to the oil. "I've had diarrhea, vomiting, the sweats, and been hospitalized for three days," Mississippi fisherman James Miller told the crowd. Miller had worked on the cleanup for more than two months.[67]

There was also a lawsuit filed by an Alabama man, Obie Carlisle, who said he became ill after trying to catch flounder with a long pole in Mobile Bay in late May. After several hours of wading in the water, Carlisle subsequently experienced "painful rashes, nosebleeds, nasal blockages, shortness of breath as well as . . . emotional distress," his lawsuit stated. Carlisle wasn't sure if he was exposed to petroleum or chemical dispersants, but both contain ingredients known to be hazardous to human health, the suit said.[68]

"There's no way you can be working in that toxic soup without getting exposures," said Hugh Kaufman, a senior policy analyst at the U.S. Environmental Protection Agency who sounded alarms early in the response to the spill about cleanup workers who were not wearing protective gear, such as respirators. Kaufman said it reminded him of the thousands of workers who went largely unprotected during the cleanup of the World Trade Center and later suffered lung disorders and other health problems. "It's unbelievable what's going on," he said. "It's like déjà vu all over again."[69]

Riki Ott, a marine toxicologist who worked for years on the response to the *Exxon Valdez* spill, said the problems in Alaska raised serious concerns about the response to the Gulf spill. "What we saw with *Exxon Valdez* was a parallel track—sick animals and sick people," she said. "Harbor seals were looking like they were drunk and dying . . . and autopsies showed brain lesions. . . . What are we exposing these poor fishermen to?"[70]

Toxicologist LuAnn White, director of the Tulane Center for Applied Environmental Public Health in New Orleans, said the worst exposures probably occurred among workers skimming oil off the surface or trying to corral the slick in booms. "When it's first released, it comes up and sits on the water," she said. "Volatile compounds in the oil, such as those that would go into gasoline or solvents—the most toxic components—evaporate on top of the water." Tar balls or "weathered oil" that makes it to the beaches is less toxic, since most of the volatile chemicals are gone, but workers should still be wearing gloves and other protection to avoid absorbing hydrocarbons, she said.[71]

The full extent of health problems was probably unknown because of the competitive nature of BP's Vessels of Opportunity program—most fishermen selected for the cleanup were unlikely to complain for fear their well-paying jobs would go to someone else. "It's an unwritten rule, you don't bite the hand that feeds you," said George Barisich, president of the United Commercial Fishermen's

Association in St. Bernard Parish, Louisiana. But Barisich said he had heard privately from many of his members about ill effects from the cleanup work.[72]

Shira Kramer, founder of a Maryland research firm called Epidemiology International that has done work for the oil industry, worried that not enough attention was being paid to health issues in the Gulf. "It's completely scientifically dishonest to pooh-pooh the potential here when you are talking about some of the most toxic chemicals that we know," she told *Bloomberg News* in June. "When you talk about community exposure, you are talking about exposures in unpredictable ways and to subpopulations that may be more highly susceptible than others, such as those of reproductive age, people who are immunocompromised, children, or fetuses.[73]

"With the World Trade Center, there have been unpredictable adverse health effects to the populations that were exposed and not just the workers," she added. "In this case, we have a soup of chemicals from the crude, chemicals from the dispersants, and pollutants that were already in the water. Who can say how they will interact?"

A grassroots group called the Louisiana Bucket Brigade, long involved in tracking health problems from refinery emissions and other oil operations in the state, enlisted volunteers throughout the summer of 2010 to conduct surveys and collect data on the impacts of the spill, including how many of the more than 30,000 people involved in the cleanup were affected. The Gulf crisis has "largely been seen as an environmental disaster," said one of the leaders of the group's effort, Shannon Dosemagen. "But the things we're tracking are the health impacts, cultural losses, economic impacts."

When the National Institute of Environmental Health Sciences

(NIEHS) announced in August that it would put up $10 million to study the health effects of the Gulf spill on response workers, NIEHS epidemiologist Dale Sandler said it was likely that data from groups like the Louisiana Bucket Brigade would serve as a starting point. "Before we do anything, we need to really engage the local community," said Sandler, who was spearheading the study with funds from the parent agency of the NIEHS, the National Institutes of Health.[74]

NIEHS Director Linda Birnbaum said the initial $10 million investment was "just a deposit" on a study that would continue for years, monitoring the long-term health effects of the spill. Though the primary focus would be on physical ailments, there were thousands—possibly even tens of thousands—of other illnesses the study would not be able to address.

Mental health counselors throughout the Gulf region reported widespread evidence that anger, depression, substance abuse, domestic strife, and, tragically, suicide were affecting individuals and households during the crisis. Some expected the problems to carry forward for months, if not years, after BP's well was capped.

"Actually, we are seeing people who are almost having flashbacks to Katrina," said Shirley Lachmann, director of five community outreach centers in Louisiana set up after the spill by Catholic Charities. "The stress level is enormously high. There are people in St. Bernard and lower Plaquemines parishes who were just getting their heads out of the water, looking forward to a great season. All of a sudden, their feet are knocked out from under them. Several are having thoughts of suicide. We have the clinical counselors down there."[75]

A federal health survey conducted door-to-door in coastal Alabama communities late in the summer found more "physically or mentally unhealthy days" for people there than in similar surveys done statewide before the spill. Many reported depression amid fears about keeping their homes or having enough money to pay for groceries. The survey of nearly three hundred households by the federal Centers for Disease Control and Prevention (CDC) also found that half the homes had at least one person experiencing respiratory problems, although it was unclear if those were connected to oil contaminants in the air.[76]

"The increased prevalence of negative quality-of-life indicators, depressive symptoms, and symptoms of anxiety suggest that resources should focus on mental health intervention and follow-up surveillance," the CDC said in its characteristically cold assessment of the findings.

The signs were clear at the five Catholic Charities centers in Louisiana, despite the fact that many fishermen tend to be fiercely independent and unwilling to openly show their feelings.

"I really believe that in this case the resilience of some of these guys actually works against them," said St. Bernard Parish president Craig Taffaro, a family counselor for two decades. "They're at the point where they start to believe they can handle anything, and when the reality starts to set in, they start thinking, 'Well, maybe we just can't handle everything. We might need some help.' And that's where the conflict begins, and that's where people begin to lose hope.

"To me, that is the most critical feature of this whole thing," Taffaro said. "If people lose hope all over again, I'm not sure that we keep them engaged. I think they run for the hills, and either drop out altogether or move away."[77]

The pastor of St. Patrick's Catholic Church in Port Sulphur, Father Gerard Stapleton, said the Gulf spill was much worse for people than the aftermath of Hurricane Katrina. "You're talking about the Vietnamese and the Croatians who have never asked for help, not from the government, not from anyone," he said. "After Katrina, they put their equipment back together. Now they've got a boat with a perfect engine with full fuel and perfect nets, and they're told they can't do anything. This is devastating for most of them."[78]

Elmore Rigamer, a psychiatrist with Catholic Charities, laid much of the blame on anger and frustration with BP. "Katrina came and it went, so how long can you be angry with God? Your anger with Katrina was more about the government's bumbling impeding your progress to recovery, and it stretched out over years. This one, it's entirely man-made," Rigamer said. "Everyone's angry at BP. Then you add the duplicity, the double-talk, the foot-dragging, and it compounds the feelings of anger."

Bonnie Duplessis, a volunteer at the Catholic Charities center in Pointe a la Hache, said working with fishermen helped her understand why her husband Ronnie, a fisherman himself, was so devastated by the spill at the start of what was expected to be one of the most productive seasons ever. "We had been suffering with so much imported shrimp flooding our market," she said. "The Louisiana Seafood Promotion and Marketing Board has been getting the word out how much better wild-caught Louisiana shrimp are for you healthwise. We were getting our highest prices since Katrina."[79]

Duplessis described her family's roller-coaster ride in an interview on National Public Radio in July. "My husband is sixty-five years old," she said. "He's been fishing since he got out of the Marine Corps. He had four kids to educate and put through college and he went to full-time fishing. And when this happened and he didn't know if he was ever going to be fishing again, it really worked on him. He's a workaholic. He has to be doing something all the time.

"BP called him, finally called our boat for the Vessels of Opportunity, and he's back to his old self because, again, he's got a purpose. I mean, even though he doesn't know if he's going to be able to fish again, at least he's able to bring in money and feel like he's supporting the family and doing something to alleviate the problem."

The work for BP was only temporary, though, and can itself be a source of anxiety, cautioned Howard Osofsky, chairman of the psychiatry department at the Louisiana State University Medical Center, appearing on the same program with Duplessis. "I've had fishermen," he said, "talk to me about how hard they would work as fishermen, they would be setting the traps for crabs. They worked very, very hard and yet were exhilarated at the end of the day. They loved looking at the Gulf. They loved their work. But now, even in working for BP, the work isn't as hard, but they come home and they're exhausted and may take a bath and go to bed."[80]

The short-term effects on mental health can be debilitating, but the long-term effects can be even more severe, Osofsky said. "And we're seeing the early signs," he said. "We're seeing anger, irritability, extreme anxiety, some conflict in families as they're dealing with the decreased economics; struggles between husbands and wives, between friends.

But it goes on and part of it is the question of what people describe as the possibility of the whole loss of life, or way of life."

Such fears apparently overwhelmed William Allen "Rookie" Kruse, a fifty-five-year-old charter boat captain from Foley, Alabama. Kruse had been plying the Gulf waters for more than forty years and earned a good living taking fishermen out in pursuit of marlin and snapper for as much as $5,000 a trip. When the excursions came to a halt after the BP spill, and his wife Tracy's seafood business started sinking, too, Kruse became depressed, lost weight, and had trouble sleeping, his wife said.

Confronted by monthly payments for his house, his pickup truck, his charter business, and other expenses, Kruse sent a pile of invoices to BP seeking help, but was doubtful about his chances of getting it. Finally in early June he was tapped for the company's Vessels of Opportunity program and worked every day on the oil spill cleanup. He hated the work, though, and became even more depressed, family members said.

On Wednesday, June 23, Kruse went to his boat early in the morning and sent his two deckhands off to run an errand before 7:30 A.M. When the crew returned, they found Kruse in the boat's wheelhouse with a bullet in his head, the handgun he kept on board for protection at his side.

"But for the oil spill, I don't think he would have done this," said his twin brother, Frank Kruse. "I guess he looked long term and saw what's going to happen when this is all over and the Gulf is dead."[81]

chapter eleven

TINY SEA HORSES

SEVERAL REPORTS IN THE LATE summer and fall of 2010 set the stage for what was likely to be years of debate about the environmental impacts of BP's oil spill on the Gulf of Mexico.

On August 4, the government almost gleefully announced that 70 percent of the estimated 172 million gallons of crude oil that had leaked into the sea had disappeared—consumed by bacteria, evaporated from the surface, skimmed by cleanup crews, or incinerated in controlled burns. While that still meant that more than 50 million gallons of oil was in the water or on the shore, the upbeat news seemed astonishing after more than three months of gloomy forecasts for the ecosystem.

"I think it is fairly safe to say that because of the environmental effects of Mother Nature, the warm waters of the Gulf, and the federal response, that many of the doomsday scenarios that were talked about and repeated a lot have not and will not come to fruition because of that," said White House Press Secretary Robert Gibbs during a briefing on the report by federal scientists.

About a month later, a coalition of scientists working on marine conservation worldwide issued a warning that a species of miniature sea horses found only in the Gulf of Mexico was threatened with extinction by the BP spill. The dwarf sea horses, barely a half-inch tall, reside in sea grasses in shallow waters that were being damaged by both the oil and the cleanup, making their survival questionable.

"We're very worried," said Amanda Vincent, director of Project Seahorse based at the University of British Columbia in Vancouver. "All of the sea horse populations in the area will be affected, but the dwarf sea horse is at greatest risk of extinction because much of its habitat has been devastated by the spill."

Two months after that, in early November, researchers studying deep-sea coral reefs near the site of the BP well found entire colonies that are usually brightly colored with waving tentacles now covered with a brown substance and virtually lifeless. "The compelling evidence that we collected constitutes a smoking gun," implicating the BP spill as the cause of the coral die-offs, said the chief scientist on the expedition, Penn State University biologist Charles Fisher.

"The circumstantial evidence is extremely strong and compelling because we have never seen anything like this—and we have seen a lot," said Fisher, a veteran of four consecutive research expeditions studying coral in the Gulf. He noted that the destruction of more than thirty coral colonies had occurred very recently and within close range of the BP well. "The proximity of the site to the disaster, the depth of the site, the clear evidence of recent impact, and the uniqueness of the observations all suggest that the impact we have found is linked to the exposure of this community to either oil, dispersant, extremely depleted oxygen, or some combination of these or other waterborne effects resulting from the spill," he said.

The disparate assessments—"This wasn't so bad after all" versus "We are destroying a rich resource"—reflected the wide range of opinions, both among scientists and in the general public, about what exactly BP had done to the Gulf with its eighty-six-day well blowout.

Some federal officials were wary of reading too much into the report that most of the oil had disappeared by August. "Mother Nature is assisting here considerably," said Jane Lubchenco, administrator of the National Oceanic and Atmospheric Administration. But, she cautioned, "Diluted and out of sight doesn't necessarily mean benign."

The point man for the cleanup, retired Coast Guard Admiral Thad Allen, also stressed that there were many problems still to be addressed. "While we would all like to see the area come back as quickly as it can, I think we all need to understand that we, at least in the history of this country, we've never put this much oil into the water," he said. "And we need to take this very seriously."

But there was also a pressing need to bolster the reeling Gulf economy after four months of endless bad news, so political leaders had a strong motive to suggest a turning point had been reached. Adding to that pressure was the fact that it was a tumultuous election year, and that President Obama and Democrats in charge of Congress were taking a beating for the sluggish economy and an unemployment rate that continued to hover around 10 percent.

"The vast majority of oil is gone," White House environmental adviser Carol Browner declared on four network news programs the morning that the report was released. The Obama administration followed up the report with announcements that Gulf seafood was safe to eat, and the president himself made a point of saying his family had no fears about consuming it. "Americans can confidently and safely enjoy Gulf seafood once again," Obama said in mid-August. "In fact, we had some yesterday."

There had been signs earlier in the summer that the damage to the Gulf might not be as catastrophic as predicted when the spill began. The NOAA reported in July that oil from the spill was biodegrading at a rapid pace and, as a result, the threat of it landing on Florida beaches was greatly reduced. There also had been concerns that heavy crude would get into the Loop Current around the southern tip of Florida and make its way up the Atlantic Coast via the Gulf Stream, but those fears seemed to be alleviated as well.

In late July, journalists flying over the spill zone reported that they saw just a few patches of sheen. Even some environmentalists were saying a significant amount of oil was being eliminated naturally: Jeffrey Short of the conservation group Oceana said as much as 40 percent of the oil on the surface of the Gulf might have evaporated.

Time magazine posed a provocative question in late July: "The BP Spill: Has the Damage Been Exaggerated?" The article by Michael Grunwald argued that only about 350 acres of wetlands had been soaked by oil, a fraction of the 15,000 acres that Louisiana was losing every year to storms, erosion, and development. Four major reasons were cited for the damage being less than what was expected: the oil from the Macondo well was lighter than the thicker crude dumped by the *Exxon Valdez;* the warm temperatures in the Gulf assisted with biodegradation; Mississippi River flows, boosted by the state of Louisiana, helped push some of the oil back into the sea; and, finally, "Mother Nature can be incredibly resilient."[1]

Coastal scientist G. Paul Kemp, a vice president of the National Audubon Society, argued in the *Time* article that omnipresent images of birds soaked in oil created a distorted picture of the spill's impacts. "There are a lot of alarmists in the bird world," Kemp said. "People see oiled pelicans and they go crazy. But this has been a disaster for people, not biota."

Official counts kept by the Coast Guard and the U.S. Fish and Wildlife Service showed that about a thousand dead animals— mostly birds, but some turtles and dolphins, too—had been collected by mid-June. By October, as the cleanup was winding down, the number of dead, oiled birds had reached 2,263, but officials were quick to point out that one hundred times as many birds had died after the *Exxon Valdez* spill in Alaska in 1989.

Although Louisiana was still building sand berms in the fall to try to block any more oil from reaching the fragile marshes on its shores, by then it was clear that the spill was doing less damage to wetlands than is done each year by storms, dredging, and coastal development, including flood-control projects. "Back of the envelope [calculations] suggest that what we're going to lose from this spill is nowhere close to the background rates of wetland loss," geologist Alex Kolker of the Louisiana Universities Marine Consortium told *Nature* magazine in September.[2]

There was still controversy in the fall about the government's declaration in August that Gulf seafood was safe to consume. Some scientists and fishermen argued that the "smell test" used by federal inspectors to determine if seafood was tainted by oil was woefully inadequate. But experts at the National Oceanic and Atmospheric Administration insisted that most fish metabolize and excrete oil fairly rapidly. "So after exposure to oil, within a matter of days or weeks, fish—and to a lesser extent shrimp and crabs and shellfish—can clear their bodies of it," said NOAA spokeswoman Christine Patrick. "If it's tainted on May 3, would it be tainted on June 3? It could pass."[3]

The U.S. Food and Drug Administration (FDA) also pointed out that samples of seafood caught by commercial fishermen were being analyzed for toxic chemicals in a laboratory, and no problems had been discovered in thousands of fish tested by fall. If a single fish was found to contain unsafe levels of contaminants, the entire area where it was caught would be closed to fishing, the FDA said.

The FDA, NOAA, the EPA, and various state agencies were all conducting tests, said the president of the Louisiana Seafood Promotion and Marketing Board, Ewell Smith. "All of them have come back with a clean bill of health, which is all different groups doing the testing," Smith said in late October. "It is the most tested food source in the world right now."[4]

A Florida toxicologist, William Sawyer, questioned whether the government tests were adequate to detect problems in shrimp, some of which had been found carrying traces of oil, and he worried that most shrimp samples were being collected in shallow waters near the shore, rather than farther out at sea where more oil was being contained by booms.[5]

Gina Solomon of the University of California, San Francisco, Medical School and a public health expert for the Natural Resources Defense Council, coauthored a paper in the *Journal of the American Medical Association* in August arguing that shrimp, crabs, and other invertebrates do absorb the polycyclic aromatic hydrocarbons, or PAHs, found in oil that are known to be harmful to human health. Solomon and her colleagues feared the PAHs would be passed up the food chain so that larger fish such as tuna would eventually carry the toxic contaminants.[6]

The head of the Louisiana Department of Wildlife and Fisheries (DWF), also said he was concerned about long-term effects from the oil. "The seafood is safe to eat, but there's a difference between consumption and the future viability of the fishery," said DWF Secretary Robert Barham. "Our fear is that some small link that is critical to the food chain could be broken."[7]

In the immediate aftermath of the spill, at least, fish populations in the Gulf seemed to be doing fine, according to both official and anecdotal assessments in the fall.

"My preliminary assessment, it looks good, it looks like we dodged a bullet," said Joel Fodrie of the University of North Carolina's Institute of Marine Science, who had been studying Gulf conditions for five years. "In terms of the numbers of baby snapper and other species present in the grass beds, things look right," Fodrie told the *Mobile Press-Register* in late September.[8]

"The fish are off the charts," said Darrell Carpenter, president of the Louisiana Charter Boat Association, in October. "There are no fewer fish. There are more fish, because they've been unharassed all summer. There are more and bigger fish. . . . The only uncertainty is all the biological science. The wild card is fish internal organs, did their eggs survive? Did they have healthy offsprings? It will take a couple of years for that to unfold."[9]

Federal scientists even reported in October that despite widespread reports of fish kills earlier in the spring and summer, there was no hard evidence that the spill was responsible for mass deaths. "In federal waters, I can tell you, there haven't been any fish kills reported that are linked to the oil spill," said Christine Patrick, the spokeswoman for the NOAA. "I know there have been fish kills reported in state waters, but I think they have determined they weren't a result of the oil spill."

Experts from the state agencies in the region concurred. "As far as wildlife, we have not observed or pinpointed any mortality in Alabama state waters of any finfish that could be attributed directly to oil," said Kevin Anson, chief biologist at the Alabama Department of Conservation and Natural Resources. "We had observed a fish kill throughout the event when there was oil in the area or offshore," he said. "But we attributed those mostly to natural phenomenon."

Of course, the government reports were sharply at odds with observations of fishermen who were hired for the cleanup and reported seeing hundreds, if not thousands, of dead fish and other animals in the waters. Riki Ott, a marine scientist from Alaska who has been working on the effects of the *Exxon Valdez* spill for two decades, spent her entire summer in the Gulf and collected reports from a network of observers about dead fish, dead turtles, even dead whales, being washed up on beaches and hauled away for disposal by response workers. "I have been down in the Gulf since

May 3. It's pretty consistent what I have heard," Ott told the *Huffington Post* in August. "First I heard from the offshore workers and the boat captains that were coming in and they would see windrows of dead things piled up on the barrier islands; turtles and birds and dolphins . . . whales." Ott said boat captains would report what she started calling "death gyres," where dead animals would be swept together by currents, but on four different occasions after precise locations were called in, the animals were gone by the time volunteers could get to the scene.

"Then I started hearing from people in Alabama a lot and the western half of Florida—a little bit in Mississippi—but mostly what was going on then, there was an attempt to keep people off the beaches, cameras off the beaches," Ott said. "Then people would—I mean you walk beaches here at night; it's hot so people walk beaches—and they would see carcasses like sea turtles, a bird, a little baby dolphin, and immediately they would go over to it and immediately people would approach them, [saying] 'Don't touch that. If you touch it you will be arrested,' and within fifteen minutes there would be a white unmarked van that would just come out of nowhere and in would go the carcass and off it would go."[10]

In fact, government agencies led by the U.S. Fish and Wildlife Service were collecting and keeping as many dead animals as they could find for the Natural Resource Damage Assessment that would help determine the amount of damage claims against BP in future litigation. Evidence-gathering began early in the spill and was expected to continue into 2011 as the Justice Department worked to develop its case against the company.

Melanie Driscoll, director of bird conservation for Audubon's

Louisiana Coastal Initiative, said part of the NRDA process would be estimating the number of birds that were killed by the oil spill but never found, as was done after the *Exxon Valdez* spill. "We do know that the count of carcasses, and the number of live, oiled birds that are rehabilitated, does not begin to enumerate the impact of the spill on populations," Driscoll wrote in October in *Audubon* magazine.[11]

Following the *Valdez* spill scientists estimated that the number of dead birds recovered was only 10 to 30 percent of the total killed by oil after the tanker accident close to the shore. The percentage of birds recovered was expected to be even lower for the Gulf spill, which began nearly fifty miles from shore, making it unlikely that many birds that died out at sea made it to land, Driscoll said.

Workers looking for dead birds in the Gulf also made a point of avoiding nesting areas until later in the summer, when many carcasses probably had disappeared, she said. "Dead, oiled birds were collected from the barrier islands at the end of the nesting season," she said. "I have spoken with some of the officials conducting the counts, and they are well aware that the bodies underrepresent the death toll. The barrier islands bake in the sun, are sometimes overwashed by waves, and may be invaded by predators. Birds that died weeks or months before the islands were searched may have decomposed beyond recognition, or been eaten by predators, or been dragged or washed away."

The bottom line is that far more pelicans and other birds native to the Gulf were affected by the spill than the 2,200 or so dead and oiled birds that were recovered. And even if the number was pushed way up by estimation methods, a body count alone does not tell the full story of possible damages from the Gulf spill.

"No list can ever do justice to what's happening in the Gulf," said Doug Inkley, a senior scientist for the National Wildlife Federation, in a July article in *Rolling Stone* magazine. "The birds that get sick and die in the wetlands will never be found. And there are so many things we are not counting. Who is out there counting the mortality among deepwater squid, which are important to the survival of sperm whales? Who is out there counting the impact on plankton, which are key to the Gulf's food chain?"[12]

The full impact on migratory birds that rely on the Gulf for food and nesting during their long journeys would not be known until at least several of the annual Christmas Bird Counts were conducted by Audubon volunteers to track changes in the numbers, said Audubon's chief scientist, Tom Bancroft. "We're going to need to pay attention to this for years to come to really understand the effects on bird populations," he said.[13]

The same could be said for many sea creatures. While it was clear that oysters in the path of the oil were quickly decimated, most fish were able to flee the areas where petroleum was present and find fresher waters. That would explain why there were no documented fish kills. But smaller forms of life, including plankton at the base of the food chain, would not have been able to escape the millions of gallons of oil and chemical dispersants dumped in the water, creating a threat that could spread through the ecosystem for years.

Scientists from the Dauphin Island Sea Lab in Alabama found

carbon signatures in zooplankton during the Gulf spill and traced the origins back to microbes that consumed oil and were in turn consumed by plankton. "We showed with little doubt that oil consumed by marine bacteria did reach the larger zooplankton that form the base of the food chain," said the lead scientist, William "Monty" Graham. "These zooplankton are an incredibly important food source for many species of fish, jellyfish, and whales."

Another study by John Kessler of Texas A&M University and David Valentine of the University of California, Santa Barbara, found widespread deaths of pyrosomes, cone-shaped filter feeders from six inches to a foot long, in areas near the BP spill. "There were thousands of these guys dead on the surface, just a mass eradication of them," said Kessler. Pyrosomes are a major food source for sea turtles and some large fish, said Laurence Madin of the Woods Hole Oceanographic Institution in Massachusetts. "If the pyrosomes are dying because they've got hydrocarbons in their tissues and then they're getting eaten by turtles, it's going to get into the turtles," Madin said.[14]

There were reports as well that traces of oil were found in both the larvae and shells of blue crabs, another food staple for fish, turtles, and birds in the Gulf. "In my forty-two years of studying crabs I've never seen this," said Harriet Perry, a biologist at the University of Southern Mississippi's Gulf Coast Research Laboratory.[15]

Oil contaminants moving through the food chain would pose a long-term threat to marine life, but there also would have been immediate impacts on larvae and eggs that happened to be in the spill zone—and BP's leak continued through the spawning season for many species. Soon after the spill began in April, teams of researchers from five Gulf Coast universities found high concentrations in the water of polycyclic aromatic hydrocarbons such as naphthalenes and dibenzothiophenes. "These are some of the most toxic components in oil," said Steven Lohrenz of the University of Southern Mississippi. The levels of PAHs were high enough—at least for a few days—to kill many small organisms such as larvae, which would have been sitting ducks in the path of an oil spill because they "are rather passive in their movement" and tend to simply float with the currents, he said. By the time the harmful chemicals had broken down or been eaten by microbes, the damage would have been done and no one would ever know about it.[16]

The nearly 2 million gallons of chemical dispersants that were dumped in the Gulf may have added to the problems for spawning animals. Peter Hodson, a marine toxicologist at Queen's University in Ontario, Canada, told a scientific conference in November that oil dispersed into tiny particles can be extremely damaging to fish eggs. "Exposures as brief as an hour can have a negative effect on embryonic fish," he said. "You could have a very large portion of the fish stock affected."[17]

The losses to various marine populations might not be known for years. "The Kemp's ridley turtles lay their eggs and the females don't return to the coast for as long as ten years," said Ronald Kendall, director of the Institute of Environmental and Human Health at Texas Tech University. "The males never come back. We won't know for years how the population is affected. The damage could well be already done."[18]

Some of the most prized fish in the Gulf could have been impacted as well, said NOAA administrator Jane Lubchenco in an

August press briefing. "For example, bluefin tuna, which spawn at this time of year, have eggs and young juvenile stages called larvae that would have been in the water column when the oil was present," she said. "If those eggs or larvae were exposed to oil, they probably would have died or been significantly impacted. And we won't see the full result of that for a number of years to come."

"Bluefin tuna spawn just south of the oil spill and they spawn only in the Gulf," agreed Larry McKinney, director of the Harte Research Institute for Gulf of Mexico Studies at Texas A&M University. "If they were to go through the area at a critical time, that's one instance where a plume could destroy a whole species."[19]

Vast underwater plumes of oil were documented by a number of scientists in the early months after the BP leak began in April. In some cases the plumes of petroleum, broken down into fine particles by dispersants, spread for miles and were in the colder, deeper waters that could make the natural breakdown of the contaminants much slower. Florida State University researcher Ian MacDonald told a congressional subcommittee in August that existence of the plumes suggested that the government's assessment that most of the oil had disappeared was "misleading." The oil "will remain potentially harmful for decades," MacDonald said. "I expect the hydrocarbon imprint of the BP discharge will be detectable in the marine environment for the rest of my life."[20]

Other scientists argued that the large volume of natural gas released from the well could also slow down the biodegradation of the oil in the sea, because hydrocarbon-consuming microbes appeared to be attracted more to the gas than the oil. "This is where [this spill] is really unique," said Rich Camilli of the Woods Hole

Oceanographic Institution, which had documented a large oil plume in June. "For most oil spills, it's just oil, it's not natural gas, but there is so much natural gas that came out of this leak. It appears as though the microbes are just interested in the natural gas. So it suggests that the oil may persist longer than we would like."[21]

Federal scientists countered in September that their studies showed the oil plumes were breaking up. "We are continuing to find lower and lower concentrations," said Sam Walker of the National Oceanic and Atmospheric Administration. "We are not finding the concentrations that they found," he said, referring to the Woods Hole research.[22]

By that time researchers from southern universities were discovering a new problem, though. While some of the oil may have been dissipating, it was also settling to the bottom of the sea. "I've collected literally hundreds of sediment cores from the Gulf of Mexico, including around this area" of the spill, said Samantha Joye of the Department of Marine Sciences at the University of Georgia. "And I have never seen anything like this. It's very fluffy and porous. And there are little tar balls in there that you can see that look like little microscopic cauliflower heads."[23]

Joye told ABC News in mid-September that a sticky, oily layer more than two inches thick in some places covered large areas of the sea floor as far as seventy miles away from BP's well. "We're finding it everywhere that we've looked," she said. "The oil is not gone. It's in places where nobody has looked for it."[24]

Another research team led by Kevin Yeager of the University of Southern Mississippi also reported finding oil on the bottom of the Gulf as far as 140 miles from the site of the well. "Clearly, there appears to be vast volumes of oil present on the sea floor," Yeager told *USA Today* in October. "We saw considerable evidence of it."[25]

The reports from the Georgia and Southern Mississippi scientists appeared to contradict earlier statements by NOAA researchers that they had looked for oil on the bottomlands, but none had been detected. "The concept of a big slick of oil sinking to the bottom is kind of an anathema," NOAA's Steve Lehmann had told the *New Orleans Times-Picayune* earlier in October. "We have not found anything that we would consider actionable at five thousand feet or five feet."[26]

But Joye said it was possible a thin layer of oil could be missed if detection equipment was dropped too abruptly on the ocean floor. "If you don't know what you're doing, you're not going to find oil," said Joye, noting that her research team had to lower its heavy "multiple corer" very gently to the ocean floor to avoid ruining the sediment samples.

Yeager said the evidence of oil on the sea bottom demanded that more research be conducted to determine the impacts on marine life. "From this point forward, this becomes largely a bottom-up story," he said. "What's troubling to me is we know almost nothing about what's happening on the seafloor in relation to this oil."[27]

Not long after Yeager made that statement, what appeared to be the first serious effects were revealed, when the Penn State researchers reported finding coral smothered by a brownish substance. "Ninety percent of forty large corals were heavily affected and showed dead and dying parts and discoloration," said a Penn State news release summarizing the findings by the team led by biologist Charles Fisher. "Another site four hundred meters away had a colony of stony coral similarly affected and partially covered with a similar brown substance."

The NOAA did take notice of the coral die-offs in a statement by its administrator, Lubchenco. "These observations capture our concern for impacts to marine life in places in the Gulf that are not easily seen," she said. "Continued, ongoing research and monitoring involving academic and government scientists are essential for comprehensive understanding of impacts to the Gulf."

The environmental group Ocean Conservancy also issued a statement saying the death of coral in the Gulf should raise the stakes for BP. "The deep-sea coral damage that has been identified by federal scientists is another reminder that the governments must vigorously explore all possible injury from this oil disaster and seek appropriate compensation in support of Gulf restoration," said the group's director of conservation science, Stan Senner. "The Gulf of Mexico is a special place, providing the nation with food, jobs, and a unique way of life. Measuring its full impact will take years, and fully restoring the Gulf will take decades."

BP was scrambling late into the fall to clean up oil-stained beaches from Louisiana to the Florida Panhandle. There were reports in October that tar balls and "weathered oil" were still landing onshore, more than three months after the Macondo well had been capped.

In Orange Beach, Alabama, which despite its name was famous for its bright white sands, cleanup workers were finding tar mats thirty inches deep in the sand, and the beach itself had a tealike tinge. Machines called "Sand Sharks" were sifting sand to remove tar balls and larger cleaning systems made by Powerscreen were attempting to scrub the sands to remove the tint. City officials had said they wanted BP to completely restore the white beaches before the 2011 tourist season, and BP spokesman Ray Melick expressed confidence the job could be done.

"I think we'll get 99 percent of what's out there," Melick told the Associated Press in November. "There may be some little BB-sized tar balls that get left, but over time nature will take care of that on its own and it will just sort of dissolve back into the surface."

"If they want me to sign off on it, it's going to have to be white and squeaky-clean," insisted Orange Beach mayor Tony Kennon. "We sell ourselves on sugar-white beaches. If we don't have that at the end of all this, we need compensation."[28]

Only one thing seemed certain about the environmental effects of the spill: they would be studied extensively for years. The disaster produced an outpouring of research dollars from public and private sources, especially BP, and larger sums were sure to follow once the government determined how much damage the oil had caused and how much BP should be made to pay for Gulf cleanup and restoration.

For starters, the company had pledged $500 million over ten years to fund scientific studies. Some of it was distributed quickly in 2010, including $10 million to the National Institutes of Health for research on human health effects from the spill and $30 million to a consortium of universities in the Gulf. But it was clear the demand would be great. When BP provided $10 million to the Florida Institute of Oceanography, made up of twenty-one research organizations in the state, more than 230 projects seeking more than $100 million were proposed, and only 27 ended up getting money. After that, the White House asked BP to give state and local officials in the Gulf a greater role in deciding how research funds should be spent, and BP agreed to funnel the money through a council of state governments in a program it called the Gulf of Mexico Research Initiative.

Federal agencies including the National Science Foundation announced funding for more than a hundred different studies, and the government also lined up scientists to help conduct its Natural Resource Damage Assessment (NRDA) for use in litigation against BP. The company, too, set about hiring consultants for the NRDA, knowing it would need expert testimony of its own to debate the damage costs in court. Dozens of researchers signed agreements to join one side or the other, but either way there was a price to pay in exchange for being hired as a consultant—all studies conducted for the NRDA must be kept strictly confidential, with no publication of data unless it was cleared for release by attorneys for the government or the defendants. "When you collect data for the [NRDA] and agree to analyze them, you are essentially foreclosing on your ability to publish those data because they're going to be involved in court cases and they're subject to all kinds of sequestering and gag orders," said Ian MacDonald at Florida State University.

The secrecy agreements scared off some researchers. BP offered a contract to the University of South Alabama's Department of Marine Sciences, but when Russ Lea, the school's vice president for research, sought to protect academic freedom, the company withdrew its offer, he said. Some scientists feared that working with BP might jeopardize their chances of getting federal research funds in the future.[29]

But other researchers concluded that government funding was declining anyway, so they jumped at the chance for more certain money from the corporation. With so much research being locked up for a legal process expected to last for years, concerns mounted that a full public accounting of what occurred in the Gulf could be delayed or compromised, especially if BP was able to suppress studies showing damages the government didn't know about. "When the best fishery scientists actually may have evidence that would work against you, but they're not able to present it to the other side or to the public, well then, you've essentially bought some silence," said Mark Davis, director of the Tulane Law School Institute on Water Resources Law and Policy.[30]

Leaders of the House Energy and Commerce Committee put BP on notice that they would be watching for "the potential suppression of scientific data and analysis" during the restoration process in the Gulf. "The disaster in the Gulf of Mexico is not a private matter," Committee Chairman Henry Waxman (D-California) and Energy Subcommittee Chairman Edward Markey (D-Massachusetts) told BP America Chairman Lamar McKay in a letter. "Mitigating the long-term impact of the oil spill will require an open exchange of scientific data and analysis."

The first steps toward a government cleanup plan were unveiled in September by Navy Secretary Ray Mabus, a former Mississippi governor who had been ordered by President Obama to set up a federal task force for that purpose. Mabus issued a report recommending that Congress establish a Gulf Coast Recovery Council that would oversee restoration efforts with money provided from any penalties that might be assessed against BP through a legal settlement with the Justice Department. While awaiting congressional action, the administration set up an interim ecosystem task force led by EPA administrator Lisa Jackson to get the cleanup effort started.

The goals for the restoration, Mabus said, should include healthy wetlands and shoreline habitats, sustainable fisheries, "adaptive and resilient" coastal communities, improved storm buffers in the region, and better management of the entire Gulf watershed and ecosystem.

Shortly after Mabus sketched out his vision of the restoration effort, leaders of environmental and other interest groups in the region met for three days at a conference center in Weeks Bay, Alabama, to develop their own set of goals for cleanup and restoration of the Gulf. The resulting "Weeks Bay Principles for Gulf Recovery" included statements that BP must be held fully accountable for damages to the Gulf and that coastal communities must again be "made whole."

"We have to make ourselves heard with one voice," said Cynthia Sarthou, executive director of Gulf Restoration Network, which organized the grassroots response to the Mabus plan. "The oil is still here, and so are we," she said, urging that the government make sure that local organizations had a seat at the table as the restoration effort was developed.

One of the world's largest conservation groups, the Nature Conservancy, also offered its support and some recommendations for the government plan, including suggestions for restoration of important bays and estuaries, better management of oil and gas development, and an emphasis on "science and technology to inform decision making" in the Gulf.

"The Gulf of Mexico is one of the few places on Earth where the health of the environment is so obviously linked to the health of the economy and community on such a vast scale," said Nature Conservancy President and CEO Mark Tercek.

"The goal should be no less than complete restoration of the bounty of the Gulf of Mexico—from its bays and estuaries to its marshes, sea grasses, fish, mangroves, coral reefs, and other plants and animals that make it one of the most biologically important and productive places on Earth."

chapter twelve

DOLLARS AND DEATHS

THE CALAMITOUS OIL SPILL IN the Gulf of Mexico hung like a mill-stone around BP's neck for the whole world to see, dragging down its reputation, its finances, and its future as a global energy giant. Yet the disaster was only the worst of BP's environmental problems in 2010, as it battled with citizens in Texas, regulators in Alaska, and lawyers in Kansas over allegations of making people sick with toxic air pollution, fouling the land through negligence, and refusing to clean up deadly hazardous wastes.

It is hard to imagine a corporation having a worse season for operations than BP had in the spring of the Macondo well blowout, and not just because of the deep-sea leak that began on April 20. At the same time the oil and gas was streaming uncontrollably into the Gulf south of New Orleans, BP's Texas City refinery was releasing more than a half-million pounds of potentially harmful chemicals, including eight and a half tons of cancer-causing benzene, into the air around the plant on the Texas coast southeast of Houston.

The leak began April 6 after a fire shut down a key component of the refinery's ultracracker unit that breaks down crude oil. When the hydrogen compressor that normally collects waste chemicals from the process failed, plant managers sent the excess fumes up a vent topped by a flare that burned off some, but not all, of the chemicals. The result was a steady stream of toxic gases spewing into the atmosphere around the refinery.

BP could have chosen to immediately shut down the ultracracker unit, but that would have meant stopping the processing of an estimated 65,000 barrels of oil per day, with each barrel worth between $5 and $10 in profits. Instead, plant managers decided to keep production going and allow the emissions to vent while the hydrogen compressor was being repaired. As a result, the leak continued for forty days, until May 16, resulting in an estimated release of 538,000 pounds of chemicals, including 17,000 pounds of benzene, into the air, state officials said later. BP informed the state on April 7 that it had an "upset" at the refinery requiring some releases beyond its permit limits, but it wasn't until June that the company reported the full volume of excess emissions.[1]

One nearby resident later told *The New York Times* that the fumes had a brutal effect on her neighborhood, although no one really knew what was happening. "We all became real sick—throwing up, diarrhea, couldn't keep anything down—and we just thought it was something that was going around," said Khristina Kelley, who lived about a half mile from the refinery with her husband and four children. "But then everybody around here got it."[2]

Mark Demark, a professor of process technology at Alvin Community College in Texas who worked at Shell Oil Company for three decades, said he understood why BP kept its unit running, but company officials also had to know they were potentially affecting thousands of nearby residents. "It's a big deal to shut the ultra-cracker down," Demark told the nonprofit news service ProPublica a couple of months after the incident. "It's operating at two to three thousand pounds of pressure, seven hundred degrees Fahrenheit, so it would take you a week just to cool that place down." But, Demark added, "Just from a public relations standpoint, for forty days to have a flare going, you have to be really inconsiderate to your community."[3]

Air pollution from the Texas City refinery had, in fact, become a harsh fact of life for the densely populated coastal area between Houston and Galveston. For more than a decade, BP had been under pressure from the U.S. Environmental Protection Agency and Texas regulators to reduce toxic emissions from the massive refinery, the third-largest in the United States. The company signed a consent agreement with the EPA in 2001 that required strict controls on benzene releases from the plant, but eight years later the EPA charged BP with failing to comply with that agreement. The company agreed on February 19, 2009, to spend another $161 million on emission controls, and it paid a $6 million fine. "BP failed to fulfill its obligations under the law, putting air quality and public health at risk," said Catherine McCabe, then acting assistant administrator for enforcement at the EPA. "Today's settlement will improve air quality for the people living in and around Texas City, many of whom come from minority and low-income backgrounds."

But the settlement with the EPA wasn't the end of the matter. Two months after it was signed, a group of refinery workers and nearby residents sued BP for toxic releases in April 2007 that put some of them in the hospital. The lawyer for the 143 plaintiffs, Patricia Davis, said the lawsuit filed in Texas court claimed the plant caused both personal injuries and a nuisance in the community. "To go there and visit," Davis said of the refinery, "you absolutely notice a difference in air quality. I was there for a week doing depositions, and at the end of the week I had a headache that wouldn't go away for days. You can smell it and can see it coming from the refineries. Who knows what residents have been breathing for years?"[4]

Texas attorney general Greg Abbott added to the pressure on BP in June 2009, citing the Texas City plant for violating state air pollution laws at least forty-six times over the previous four years, and seeking more than $100 million in penalties. "BP Products in Texas City has become an environmental disaster," Abbott said. "It's time that BP Products cleans up its act."[5]

The state action was the last straw for a group of BP investors who filed suit in August 2009 charging the company's management with hurting earnings by failing to comply with environmental laws. "The intentional or reckless misconduct has caused, and continues to cause, BP significant damage from various claims, lawsuits, investigations, enforcement proceedings, as well as the associated fees and penalties," the shareholders alleged in a complaint to the state court in Galveston County.[6]

More workers at the refinery also went to court saying emissions from the plant were making them sick. After hearing one of the cases in December 2009, a federal jury in Galveston recommended that ten of them be awarded $10 million each for being exposed to hazardous emissions at the refinery in 2005. The $100 million verdict infuriated BP, which said the incident in question was merely an "odor event" and that it was "outraged" by the jury verdict. But the workers' attorney, Tony Buzbee, said the company was in denial about its pollution problems. "They're like an ostrich with its head in the sand," Buzbee said. "They don't understand the meaning of responsibility."

Buzbee said he initially asked for only $5,000 for each worker, but when BP offered just $500 to each, he took the case to trial. When the arguments concluded, the judge told the jurors that if they believed "BP's conduct was so shocking and offensive as to justify an award of punitive damage," they should consider making one. But even Judge Kenneth Hoyt was apparently stunned by how far the jury went in that direction; a few months later he knocked down the award for the ten workers to a total of $340,660, saying there was not enough evidence that BP's conduct had been grossly negligent to allow the $100 million award to stand. Nonetheless, the point had been made that some in the community were fed up with emissions from the BP plant.[7]

Six more workers went to the state court in Galveston in January 2010 seeking $500 million in damages after being exposed to vapors when a pipe broke at the refinery five months earlier. The employees had been told while being treated at the hospital that it appeared they had been exposed to benzene.[8]

And when it was reported in the summer of 2010 that the refinery had unleashed a steady stream of benzene and other chemicals in April and May, attorney Buzbee went into action against BP again. This time he found a very sour mood among residents of the Texas City area who had long been generally supportive of BP because of its enormous contribution to the local economy. More than two thousand people signed on for a class action lawsuit demanding $10 billion in damages from BP, and several thousand more lined up at a convention center the day after the suit was filed seeking to add their names as plaintiffs.

"I've never seen anything like this," Buzbee told *The Houston Chronicle* as he watched people file past to meet with a team of lawyers helping with the case. "I can't believe this is mass hysteria and that everybody here is a faker." A former nurse, Linda Laver, told the *Chronicle* that she never had serious health problems in all her fifty-five years until she came down with sinus problems, pneumonia, and gallbladder failure during the weeks of the leak. Laver said she "freaked out" when she learned about the benzene emissions from an ad placed by Buzbee's law firm. "It confirmed all my suspicions," she said.[9]

The Texas attorney general also came down hard on BP for the April and May emissions. This time, Abbott charged the company with violating state environmental laws in order to avoid a shutdown that would have cut into its profits. "BP decided to continue those units so as not to reduce productivity," his complaint said. "BP made very little attempt to minimize the emission of air contaminants caused by its actions, once again prioritizing profits over environmental compliance." The EPA entered the fray as well, announcing in September that it would open a criminal investigation into chemical releases at the refinery. "By joining the investigation, the EPA will help ensure disclosure of all information by BP," said the EPA regional administrator, Al Armendariz, in a statement about the probe. "It is important the EPA, state officials, and public know what happened at the plant, and that BP is held accountable to prevent incidents like these from happening in the future."

Around the same time BP was battling regulators and Texas City residents over its air pollution, the federal Occupational Safety and Health Administration was hammering the company for safety issues inside the refinery. OSHA announced in October 2009 that BP had still failed to correct serious problems identified after fifteen workers were killed at the refinery in March 2005, and proposed a fine of $87.4 million, the largest in the agency's history. "This administration will not tolerate disregard of our laws," said Labor Secretary Hilda Solis in announcing the enforcement action by her agency, adding that if BP continued to violate safety standards at the refinery there could be another terrible accident. "BP still lacks important information about the hazards of its pressure vessels and the company continues to lack clear operating procedures that could prevent another disastrous explosion," Solis said.

BP contested the fine with OSHA's review board, insisting it had abided by its agreement to address safety issues and had spent hundreds of millions of dollars on plant improvements after the tragedy in 2005. "We continue to believe we are in full compliance with the settlement agreement," said refinery manager Keith Casey, "and we look forward to demonstrating that before the Review Commission." He added in a statement to the *Galveston Daily News,* "Let's be perfectly clear, if we believed the refinery was unsafe, or any unit in the refinery was unsafe, we would take immediate corrective action including, if necessary, shutting it down."[10]

But less than a year later, BP agreed to pay $50.6 million to settle the OSHA complaint. The penalty, though reduced from the $87.4 million originally proposed, was the largest ever assessed in the agency's history. It was more than double the previous record—a $21 million fine levied against BP after the March 2005 explosion that killed 15 workers and injured 180 others.

In March 2010, on the five-year anniversary of the Texas tragedy, the plant manager in Texas City, Casey, said the accident had "fundamentally changed BP" and led to more than $1 billion worth of upgrades at the refinery. At least one official of the federal agency that investigated the accident five years earlier wasn't sure how much the lessons had sunk in at the company. "When will we know whether the tragedy of 2005 has resulted in greater safety at BP and other companies' refineries?" John Bresland, chairman of the Chemical Safety Board, asked in a statement on the anniversary of the accident. "Only when we can look back over the passing of a significant number of years without major accidents, deaths, or injuries."

In May 2010, investigative journalists for the nonprofit Center for Public Integrity in Washington released a startling finding about BP's refinery operations. Fully 97 percent of all "flagrant violations" found throughout the entire U.S. refinery industry by OSHA in the previous three years had occurred at just two BP refineries, in Texas City and Toledo, Ohio. "The only thing you can conclude is that BP has a serious, systemic safety problem in their company," the U.S. Labor Department's deputy assistant secretary for occupational safety and health, Jordan Barab, told the center.[11]

Barab reiterated his concerns a month later at a Senate subcommittee hearing on safety issues in the energy industry in the wake of the eleven deaths on the *Deepwater Horizon*. Barab linked the tragedy in the Gulf to BP's ongoing safety problems at its refineries, noting that OSHA had proposed the record fine of more than $87 million the year before at Texas City because BP "had failed to abate many of the problems that it agreed to address after fifteen workers were killed in the 2005 explosion." OSHA also hit BP's Toledo refinery with $3 million in penalties for "egregious willful violations" in March 2010, five years after the company's tragic accident in Texas. "This failure to learn from earlier mishaps has exacted an alarming toll in human lives and suffering," Barab told the Senate panel.

BP's problems with safety and environmental regulators extended north to Alaska, where a growing number of equipment failures and leaks were being reported even after the company spent millions of dollars to upgrade its pipelines after corrosion problems forced a nearly two-month shutdown of the nation's largest oil field at Prudhoe Bay in 2006. A break in a BP pipeline in March of that year had dumped almost 200,000 gallons of oil on the North Slope—the largest land spill in Alaska history—and exposed many weaknesses in the feeder system that pumps thousands of barrels of oil each day into the Trans-Alaska Pipeline System for transport to the lower forty-eight states.

Serious problems were still cropping up years later, enough to prompt the chairman of a congressional panel investigating BP to say there was greater risk of a disaster at Prudhoe Bay than there was of another big spill occurring in the Gulf of Mexico. "I'm more concerned about Alaska, and no one seems to be paying attention," said Representative Bart Stupak (D-Michigan), who chaired the

House Energy and Commerce oversight subcommittee that investigated BP's record in both Alaska and the Gulf. "We're just one major accident from shutting down our most strategic pipeline," Stupak said in an interview before retiring from Congress at the end of 2010.

Although oil output from Prudhoe Bay has dropped about two-thirds from its peak production of 1.5 million barrels a day in 1988, it remains the nation's biggest source of domestic petroleum. And as the wells are tapped down, they tend to produce a more corrosive mixture of oil, water, and mineral deposits, making proper maintenance of the vast pipeline system in Alaska more important than ever.

But Stupak's investigators in 2010 found there had been a series of "significant events" in BP's oil and gas operations in the years following the 2006 Prudhoe Bay shutdown, any one of which could have caused another production stoppage, along with worker injuries or deaths and serious environmental damage. For instance, in September 2008 a BP pipeline carrying highly pressurized gas from a Prudhoe Bay drilling site ruptured, sending two large sections of pipe flying more than nine hundred feet through the air. A dozen workers in the vicinity avoided being hit, but said they felt vibrations from the ground when the accident occurred. A state report said later that "had the high-pressure gas pipeline failure occurred under slightly different circumstances, the results would have been catastrophic, potentially with loss of life."[12]

Another leak caused by a pressure buildup, possibly from an ice blockage, occurred in a BP oil pipeline on the North Slope in November 2009, dumping an estimated 46,000 gallons of mixed crude oil and water that contaminated ice and snow in the area. It was the second-biggest land spill in Alaska history—exceeded only by the BP leak that resulted in the Prudhoe Bay shutdown in 2006—and prompted the EPA and the FBI to open a joint criminal investigation to determine how a two-foot gash opened in a critical pipeline. A year later, in November 2010, a federal probation officer overseeing BP operations in Alaska, Mary Frances Barnes, went to federal court to argue that the company should face criminal negligence charges for the spill. The incident was still under investigation at the end of the year.[13]

About the same time in late 2010, investigative reporters at the nonprofit news service ProPublica uncovered an internal BP document that showed 148 of its pipelines in Alaska had received an "F-rank" grade, meaning that more than 80 percent of the pipeline walls were significantly corroded.[14] Workers on the lines in Alaska were also expressing grave concerns about the state of BP's oil and gas delivery system in 2010. "The condition of the [Prudhoe Bay] field is a lot worse and in my opinion a lot more dangerous," said a leader of the United Steelworkers union, Marc Kovac. "We still have hundreds of miles of rotting pipe ready to break that needs to be replaced. We are totally unprepared for a large spill," Kovac, who spent three decades working with BP, told the investigative news service Truthout. Kovac said many other workers were concerned but afraid to risk their jobs by speaking out; he did so, he said, because "many lives are at stake."[15]

BP remained under heavy pressure from state and federal agencies concerned that workers, the environment, and the nation's energy supplies were threatened by the company's management problems in Alaska. Both the State of Alaska and the U.S. Justice

Department filed civil suits against BP on the same day in 2009, charging violations of environmental laws. The state's lawsuit also demanded that BP reimburse Alaskans as much as $1 billion for revenues lost while the Prudhoe Bay field was shut down in 2006, although a judge later limited BP's liability to around $300 million. The cases were still pending in late 2010, despite several attempts by BP attorneys to force dismissal.

A $10 BILLION DAMAGE SUIT in Texas and a multimillion-dollar reimbursement demand from the State of Alaska paled by comparison, of course, to BP's potential liabilities in the Gulf of Mexico. But added together, the company was facing penalties, cleanup costs, and damage claims totaling anywhere from $50 billion to $100 billion as it entered 2011. It seemed unlikely that BP's costs in the Gulf would exceed the record for legal settlements set by the tobacco industry with its agreement to pay $206 billion for health claims in 1998, but they would certainly place the company second on the list of all-time payouts for corporate wrongdoing, far surpassing the $7.2 billion that Enron Corporation paid for defrauding investors in 2006.

BP's own estimate in November was that the Gulf spill would cost the company about $40 billion, but some analysts predicted the total tab would be much higher. There was already $20 billion committed to the victims' compensation fund that was being distributed by Washington attorney Kenneth Feinberg; thousands of claims were still being considered in late 2010. Penalties and damages assessed by the government against BP would very likely be the most ever demanded in an environmental case, with the possibility that fines under the Clean Water Act alone could exceed $17 billion if BP was found grossly negligent in causing the spill. Then there was the possibility of punitive damages being awarded to victims of the accident, especially to the families of the eleven men who died on the *Deepwater Horizon*, which could escalate settlement costs well into the tens of billions of dollars.

How much BP would ultimately pay for the Gulf spill would largely be determined by the work of the "BP squad," a team of more than forty U.S. attorneys and scores of investigators putting together the federal case against the companies and individuals involved in the disaster. Operating out of an office in New Orleans, and bolstered by support from Justice Department headquarters in Washington and other federal agencies, the squad was expected to work well into 2011 developing a complaint that, barring a settlement, probably would not be argued in court until later in the year at the earliest.

The government team was focused on possible violations of two major laws—the Oil Pollution Act passed after the *Exxon Valdez* accident to spell out liabilities for spills, and the Clean Water Act that prohibits damage to the nation's waters—but it could also look for breaches of the Endangered Species Act, the Migratory Bird Treaty Act, the Marine Mammal Protection Act, and other statutes to compile possible charges in the case. A Natural Resources Damage Assessment being conducted by federal agencies and contractors would help determine the value of Gulf assets that were damaged in the spill, such as the number of pelicans or turtles that were killed, oyster beds that were wiped out, and wetlands that were destroyed.

If they had enough evidence that BP managers engaged in a

companywide pattern of behavior aimed at skirting federal laws and regulations, U.S. attorneys could also utilize a law enacted in 1970 to give prosecutors a powerful tool to use against organized crime, the Racketeer Influenced and Corrupt Organizations statute or RICO. Several attorneys for businesses claiming damages from the spill filed suits built around RICO, alleging that BP executives conspired to deceive the government about the company's ability to drill for oil safely in the deep waters of the Gulf. A lawsuit filed on behalf of Florida property owners by attorney J. Michael Papantonio claimed that BP knowingly submitted false reports to the Minerals Management Service that understated the risks of a deepwater well blowout and misrepresented the company's ability to respond to an accident.[16] Another lawsuit by Louisiana attorney Daniel Becnel seeking damages for restaurants and other businesses hurt by the spill also argued that BP violated RICO by filing "false documents" with the government, and even bribed regulators with gifts, alcohol, drugs, and sex to get drilling permits rubber-stamped. "The pattern of racketeering activity engaged in by defendants involves a scheme to fraudulently create a pretense of safety to the public while, at every turn, seeking to avoid the costs associated with actually conducting their operations in a safe manner," Becnel's suit alleged.[17]

BP's angriest critics argued that the company deserved to be treated as a kind of corporate mafia that put profits above human life in its operations in Texas and the Gulf, but citing the RICO statute would be a devastating blow to the biggest producer of oil in the United States. President Obama, despite sharply criticizing BP during the height of the Gulf spill, had stressed that the government was not out to destroy the company, but only wanted to make certain it paid for all the damages that it had done, and BP executives had publicly pledged to do so.

The government had a strong motive, though, to prove that BP was reckless or took excessive risks in its operations leading up to the blowout. The Clean Water Act spells out fines of up to $1,100 for each barrel of oil dumped into a U.S. waterway, but the penalty can go up to $4,300 per barrel if a polluter is found guilty of gross negligence in causing a spill. Based on the government estimate that 4.1 million barrels of oil leaked into the Gulf from the Macondo well, the fines against BP under the law could range from $4.5 billion to $17.6 billion, depending on how much negligence could be proved. The difference would be especially significant if the government was able to direct most of the penalties ultimately assessed against BP into a cleanup fund for the Gulf, as recommended by Navy Secretary Ray Mabus, tasked by President Obama to develop a Gulf restoration strategy.

Federal investigations of the spill would not be completed until 2011, but there were already indications by the end of 2010 that a number of them would cite BP and its partners on the *Deepwater Horizon* for not taking adequate steps to ensure against a well blowout.

Engineers and scientists who investigated the spill for the U.S. Department of the Interior planned to issue a final report in June 2011, but a preliminary report on their findings in November 2010 said the drilling in the Macondo field was conducted with "insufficient consideration of risk." The expert panel from the National Academy of Engineering and the National Research Council also said there was evidence that BP was trying to save money by rushing to complete the well and seal it up for production later. "Many of the pivotal choices made for the drilling operation and temporary abandonment of the well were likely to result in less cost and less time relative to other options," the technical panel's interim report said.

The investigators went on to cite numerous instances in which proper precautions were not taken on the drilling rig, including decisions to ignore pressure tests showing there could be gas and oil leaking into the well, to use fewer centralizers on the well pipe than recommended to ensure the cement casing did not leave room for gas to push through, and to skip a final, time-consuming test called a "bond log" that would have determined the integrity of the cement job in the well. The expert panel also criticized BP for choosing a less expensive, but faster method of building the well that used only a single long string of pipe rather than a dual-layered well with a cement liner to provide an extra level of protection from a blowout.

After several months of hearings on the accident, and in the final months before submitting a report to the White House on the spill, leaders of the presidential commission investigating the BP disaster also raised questions about risk-taking during the drilling operation. "The series of decisions that doomed Macondo evidenced a failure of management, and good management could have avoided a catastrophe," said William Reilly, cochairman of the National Commission on the BP Deepwater Horizon Oil Spill and Offshore Drilling, at one of the panel's final hearings.

Reilly, a former EPA administrator under President George H. W. Bush, issued a harsh assessment in December of the performance by managers from Macondo well owner BP, *Deepwater Horizon* owner Transocean, and cement contractor Halliburton. "There is virtual consensus among all the sophisticated observers of this debacle that three of the leading players in the industry made a series of missteps, miscalculations, and miscommunications that were breathtakingly inept and largely preventable," he said.

The commission's other cochairman, former senator Bob Graham (D-Florida), also blamed the disaster on a lax safety effort by BP and its contractors. "The problem here is that there was a culture that did not promote safety, and that culture failed," Graham said at that same hearing.

It remained to be seen whether ineptness could be translated into gross negligence in the conduct of the drilling operation, but the commission also uncovered evidence that the cement mixture used to seal the well did not pass lab tests for strength and stability, and that managers for Halliburton—and possibly BP—signed off on it anyway. Critics of BP and its partners jumped on the finding as proof of their negligence. "The fact that BP and Halliburton knew this cement job could fail only solidifies their liability and responsibility for this disaster," said Representative Edward Markey (D-Massachusetts). "This is like building a car when you know the brakes could fail, but you sell the cars anyway."

BP's own investigation of the spill, released in September 2010, not surprisingly tried to place a large share of the blame for the accident on Halliburton and Transocean. The company's 193-page report on the events leading up to the blowout listed eight "findings of fault," and only one put blame on BP's managers along with those from Transocean—for misinterpreting pressure tests on the well before allowing the drilling mud to be removed, a decision the company later admitted was disastrous.

"No single factor caused the Macondo well tragedy," BP insisted in its report. "Rather, a sequence of failures involving a number of different parties led to the explosion and fire which killed eleven people and caused widespread pollution in the Gulf of Mexico earlier this year." In particular, the company laid blame on Halliburton for a cement job that "failed to contain hydrocarbons within the reservoir, as they were designed to do, and allowed gas and liquids to flow up the production casing."

Transocean and Halliburton responded that every decision and every test conducted during the drilling operation was ultimately approved by BP. "The well owner is responsible for designing the well program and any testing related to the well," Halliburton said in a news release responding to BP's report. "Contractors do not specify well design or make decisions regarding testing procedures as that responsibility lies with the well owner."

The finger-pointing between the companies was a delight to attorneys for victims of the spill, since it indicated those responsible for the accident would not be presenting a united front in court battles to come. The accusations flying among the defendants could also enhance the prospects of winning settlements from all the companies, not just from BP, plaintiffs' lawyers said. The potential problems revealed about Halliburton's cement, for instance, were seen as a "game changer" by Tony Buzbee, an attorney representing workers who were injured on the *Deepwater Horizon* as well as hundreds of businesses that claimed they were harmed by the resulting oil spill. "The findings, of course, further solidify BP's liability, but now put Halliburton squarely in the gun sights of thousands of lawsuits," Buzbee told the *Financial Times*.[18]

Evidence that BP and its partners took shortcuts that led to the well blowout would be pivotal not only for the criminal investigation, but also for plaintiffs in civil suits seeking punitive damages against BP. First in line among them were the families of the eleven men who died on the *Deepwater Horizon*, all of whom were likely to seek far more compensation from BP and its contractors than the lost wages of the victims, the maximum allowed under an outdated federal law covering shipping accidents at sea. While Congress was moving toward amending the law to allow all forms of compensation, including punitive damage awards, for victims of shipwrecks, lawyers for the victims' families were already building their cases for millions of dollars in settlements from BP, assuming none of the cases would actually go to trial.

"BP is not going to let the death cases go to trial," said Keith Jones, a Louisiana attorney whose son, Gordon Jones, died on the *Deepwater Horizon*. At the same time, he said, "BP is mighty exposed, whether damages are based on the extent of their ability to pay or on the size of their wrongdoing. The issue that BP fears is not paying for the damages they caused—they've got lots of money for that. What they fear is punitive damages."

If the company is found guilty of gross negligence for the spill, Jones said, "they could be forced to pay huge punitive damages." For his part, Jones said he knew that "all the money in the world" would not make up for the loss of his twenty-eight-year-old son, Gordon. But when he sat down at the table with BP to demand adequate compensation for Gordon's widow and her two young children, he said he wanted to ensure that the family "would never have to worry about money for the rest of their lives."

While substantial settlements seemed almost a certainty for the survivors of those killed on the rig, the prospects were less clear for both defendants and plaintiffs heading down the path of civil litigation that would likely last for years. For BP and its codefendants, the sky could be the limit on their costs if the courts awarded punitive damages to many of the thousands of plaintiffs demanding billions of dollars from the companies. For the plaintiffs, the only guarantee was they would be facing fierce attacks from some of the finest corporate attorneys that money could buy, and the battles could last for years if BP, as would be expected, appealed any unfavorable decisions.

Stuart Smith, a New Orleans attorney representing fishermen and businesses hurt by the spill, said he knew that the mother of all legal battles was in store for plaintiffs. "I've been suing oil companies for pollution almost exclusively for twenty-three years," Smith told *The Washington Post*. "And oil companies are the meanest, nastiest defendants in the country. They just don't care; they have so much money."[19]

While the road ahead was daunting, there was one positive sign. Most of the civil cases were consolidated in the federal court in New Orleans, to be heard by U.S. District Judge Carl Barbier, widely considered to be one of the fairest judges in the federal system in Louisiana. "Carl Barbier is good news," said Keith Jones. "There isn't going to be twenty-year litigation like with the *Exxon Valdez*. Judge Barbier won't allow that. Our concern is with the Fifth Circuit and the Supreme Court of the United States." The appellate courts tend to be less friendly to plaintiffs, he said, as was seen in the *Exxon Valdez* case that dragged on for two decades and saw a $2.5 billion damages award to victims reduced to about $500 million in the process.

The federal appellate court in Louisiana also could be an obstacle for plaintiffs in spill cases, according to Nan Aron, president of the Alliance for Justice, a Washington group that advocates for legal rights of victims and minorities. BP could move forward with appeals knowing that fifteen of the twenty-one judges on the U.S. Court of Appeals for the Fifth Circuit in the Gulf region had "huge financial holdings in oil and gas" that could influence their decisions about damage awards against energy companies, Aron said.

BP insisted throughout the crisis in the Gulf that it would pay all "legitimate" damage claims from the spill, and the company's financial stature indicated it should be able to make good on that promise. After all, it was a global energy corporation with 18 billion barrels of proven oil reserves in its portfolio, along with lease holdings that had the potential to produce 63 billion more barrels. The company's total market value was approaching $200 billion before the spill and it had operations worldwide producing on the order of $30 billion in revenues every year. Financial analysts noted that BP had only $26 billion in debt at the end of 2009, for a debt-to-equity ratio of less than 20 percent, a level that should allow it to easily obtain loans for spill expenses if needed. "Our asset base is strong and valuable," BP said in a statement at the height of the spill in June, with "significant capacity and flexibility in dealing with the cost of responding to the incident, the environmental remediation, and the payment of legitimate claims."

BP did take a beating on its stock value virtually every week that the spill continued, to the point that its total market value was cut in half in June. The losses on paper gave rise to speculation about a possible takeover, perhaps by U.S. energy giant ExxonMobil, but BP took steps to ward off any such thoughts in rival corporate boardrooms. A source within the United Arab Emirates told Reuters in July that BP had approached sovereign wealth funds in Abu Dhabi, Kuwait, Qatar, and Singapore seeking "a strategic partner" to help defend against a takeover bid.[20]

There was an occasional mention of possible bankruptcy for BP during the early weeks of the spill, perhaps fueled by a statement from Interior Secretary Ken Salazar, standing outside BP headquarters in Houston, that the company's "life is very much on the line here." ABC News reported that the Federal Reserve Bank of New York surveyed major banks about their BP holdings to ensure there would be no risk to the financial system if the company collapsed, but a source said the central bank determined that none of the banks was overexposed.[21]

Late in June, as public fury raged in the United States over BP's failure to cap the well, British Prime Minister David Cameron told the Canadian Broadcasting Corporation that he feared the company would be destroyed. "I think it is also in all our long-term interests that there is some clarity, some finality, to all of this, so that we don't at the same time see the destruction of a company that is important for all our interests," Cameron said on the CBC. "This is a vital company for all of our interests. The view I take is that BP itself wants to cap the well and clean up the spill and compensate those who have had damages. It wants to do these things, it will do these things. I want to work with everyone concerned to try to make sure that out of all this there will still be a strong and stable BP, because it is an important company for all of us."[22]

For the most part, though, BP executives expressed confidence about the company's ability to survive the crisis in good health, with some even proposing to pay about $10 billion in dividends to shareholders while the leak was still raging in the summer. They dropped the idea under pressure from U.S. politicians, including President Obama.

After one of the worst three-month periods in its one-hundred-year history, BP reported a net profit of $1.8 billion in the third quarter of 2010—a significant drop from the $5.3 billion earned in the third quarter of 2009, but a profit nonetheless. The company did make a move considered prudent by investors when it announced in July that it would put $30 billion to $40 billion worth of assets on the market to raise cash for future spill expenses. Although it would mark a sudden shift from the aggressive expansion launched by former CEO John Browne in the 1990s, the proposed sales would not diminish the company's future prospects for growth. Most of the assets to be sold were older or shallow-water wells with declining output, including some on Prudhoe Bay in Alaska. Sales also were targeted for natural gas fields in the western United States and Canada and for smaller blocks of assets overseas, including some in Venezuela, Vietnam, and Pakistan. What the company intended to keep were the holdings that got it into trouble in the first place, the operations that were at once the most lucrative and the most risky for the company, the deepwater drilling programs in the Gulf and elsewhere around the world.

BP, said the company's new CEO, Robert Dudley, in November, "is now going to become incredibly focused on managing the risks, for example, of deep water. It's not going to shy away from the risk, it's going to get even better at it." In an interview with *Bloomberg News,* Dudley admitted it would take time to get those operations back on track in the wake of the spill. "Before we go back to the work in the Gulf, even if the government said it's okay to go back now, we are going to go through all of our operations and rigs and working processes with great, great detail," he said. "We're not going to rush back into the Gulf."[23]

At the same time, Dudley made it clear that BP had no intention of abandoning its operations in the United States, no matter how much pressure came from regulators, litigators, and the public at large. "I can promise you that I did not become chief executive of BP in order to walk away from the U.S.," Dudley said in a fiery speech to the Confederation of British Industry on October 25 in Houston. "BP will not be quitting America."

Dudley suggested that BP now defined Big Oil in the United States, having taken over many of the offspring of the Standard Oil Company that dominated the industry until it was broken up by a historic antitrust case a century before. "BP's heritage companies such as Amoco, ARCO, and Sohio can trace their roots back many decades, in some cases to the 1860s and '70s," Dudley said. "Today as BP, we are the leading producer of oil and natural gas in the United States. We currently employ around 23,000 people directly in this country and we support around 200,000 further jobs. We have 75,000 retirees. They live in all fifty states. We have a half-million individual sharcholders. We have over $55 billion in operating capital employed, including five refineries, and we sell more than 15 billion gallons of gasoline here every year."

Dudley drew the line on reopening the Macondo oil field to drilling, however, despite its potential to produce millions of barrels of oil and gas. "The well will be capped, and nobody will ever want to go near that well again, and there are no plans to develop that Macondo structure field," he had said even before he took over as CEO in October.[24]

While digging in deeper in the United States, BP also was taking steps to expand operations in Canada that could insure its future as North America's top oil producer. In November 2010, BP announced it was moving ahead on its joint venture with Canada's Husky Energy to start production from the oil-rich sands of Alberta by 2014. Experts believed there could be as much as 175 billion barrels of recoverable petroleum locked in the soils of the western province, an amount surpassed only by the abundant reserves in Saudi Arabia. Earlier in 2010, the highly respected Cambridge Energy Research Associates projected that by 2030, the Canadian oil sands could become the source of more than a third of America's petroleum imports.[25]

The downside was that the technology for extracting oil from the sands posed enormous threats to land, water, and wildlife, not to mention the added impacts on climate change from expanding the production and use of fossil fuels. And so it seemed that one of the results of the disaster in the Gulf of Mexico was the clear return of BP to the "Big Oil" fold it had shunned more than a decade earlier when it pledged to tackle global warming and promised the world that it was a company moving "Beyond Petroleum."

"I don't think anybody with a straight face is going to say that they're an environmental leader for quite a while," said Andrew Winston, coauthor of the book *Green to Gold: How Smart Companies Use Environmental Strategy to Innovate, Create Value, and Build Competitive Advantage* and a consultant to businesses on improving both environmental and financial performance. "There would have to be a pretty significant shift. It's not just the spills and safety [issues], they actually reduced some of their investments in renewable energy and basically went in the opposite direction and just went back to being a petrochemical company."

BP's former CEO John Browne was admittedly a pioneer in green marketing, but after his departure in 2007 the company seemed to abandon its focus on environmental management, Winston said. "I think what happened at BP, it seems like different leaders came in after Browne, and they made cost-cutting the goal, and pure profitability. And a lot of practices they should have been keeping an eye on fell by the wayside," including worker safety and environmental protection, he said. It was unlikely BP could ever recover its green image, but that shouldn't be its goal anyway, Winston said. "Once you're in a crisis management, there's not a lot of brand play, you just need to be doing the right thing," he said.

"How much are they going to be helping people rebuild, providing money for areas that were damaged, or tourism that was damaged, or how much are they going to be fighting?" Winston asked. "And are they putting in place policies that stop the kind of accidents that kill their own employees? They've had loss of life multiple times now. It's really not acceptable."

BP's new CEO was promising a complete overhaul of the company's approach to safety, much as his predecessors Browne and Tony Hayward had pledged after the tragedy in Texas City. In announcing new efforts to reward employee performance in risk management and safety, Dudley said, "These are the first and most urgent steps in a program I am putting in place to rebuild trust in BP—the trust of our customers, of governments, of our employees, and of the world at large." But his words rang hollow for many Americans, after months in which BP officials seemed to be downplaying the extent of the oil leak in the Gulf, pointing fingers at its former partners for causing the accident, and making statements seen as insensitive toward the plight of people affected by the tragedy.

"This has to be one of the all-time disasters for corporate reputation," James Hoopes, a professor of business ethics at Babson College in Massachusetts, told *The Guardian* newspaper in London. "The most graceful course of action for BP would be to hang its head for a very long time and admit it has some deep issues to deal with."[26]

BART STUPAK, WHO LED INVESTIGATIONS of BP's problems in Alaska and the Gulf before leaving Congress at the end of 2010, said that as a holder of leases in the Gulf, BP had a right to seek new drilling permits. But he said the U.S. government's response should be: "Go ahead and apply, but you've not demonstrated you can do it safely." On the Macondo well, Stupak said, "There's no doubt that they cut corners." And in Texas City, the company had more "willful violations" of OSHA requirements at its refinery than any other plant in history, he said.

372] POISONED LEGACY

Stupak said he could never forget the fifteen workers who died at the BP refinery explosion in Texas and the eleven more who were killed while drilling the BP well in the Gulf. "When do you stop this madness?" he asked.

Dudley acknowledged in his October 25 speech in Houston that BP had a long way to go to rebuild trust with the public. "When people look at a BP logo on a gas station today they probably associate it with the accident and the spill," he said. "But our actions can help inform perceptions. And I would hope that people are starting to think about the magnitude and intensity of our response to that spill and the way we are doing as much as we can to restore livelihoods, look after the environment, and rebuild relationships.

"I would hope they would also see a company that has suffered a terrible accident but has the humility and courage to learn from that incident and prevent such a thing happening again. I'd hope they would also see a company that is determined to do the right thing by the people of the Gulf region and across the United States. I prefer to look our customers in the eye and say to them, 'We're sorry about what happened, but we're not running away and we're going to make it right.'"

Two days after its CEO spoke those words, BP was doing all it could to run away from a toxic waste site in the heart of America that was the legacy of the very "heritage companies" Dudley had proclaimed as his own.

ON OCTOBER 27, 2010, THE Kansas Supreme Court heard oral arguments in the case of *Neodesha v. BP,* the lawsuit filed more than six years earlier by the southeast Kansas town seeking a cleanup of petroleum wastes that BP had inherited from Standard Oil and its successor Amoco, which operated a refinery in Neodesha from 1897 to 1970. It had been nearly three years since a jury, following what most legal observers believed to be the longest civil trial in Kansas history, had exonerated BP from any further responsibility for cleaning up the abandoned refinery site, and it was just more than two years since Kansas District Judge Daniel Creitz tossed out the jury's verdict and ruled that BP was, in fact, liable for damages in the town. "Given the defendants' admissions and evidence here, no rational juror could return a verdict stating that [BP was] not guilty of contaminating the groundwater underneath Neodesha," Creitz said in his ruling. "The contaminants in this case are some of the most dangerous known to mankind."

Not much had happened in Neodesha since Creitz's September 2008 decision reversing the ten "not guilty" verdicts the jury had issued at the beginning of that year. The Kansas Department of Health and Environment (KDHE) completed a "Community Involvement Plan" for the abandoned refinery site that BP had acquired when it bought Amoco in 1998 and found thick deposits of petroleum wastes sitting on top of the groundwater plume that stretched beneath the entire city between the Fall and Verdigris rivers. BP had agreed to pump out some of the pollution in hopes of reducing the level of contaminants, but since the community tapped the Verdigris for its water supplies, the company felt the waste posed no threat to the public and argued the cleanup could best be handled by Mother Nature, even though the "natural remediation" could take more than a hundred years. The KDHE had gone along with BP's strategy, and set up the

Community Involvement Plan in October 2008 "to ensure that residents are informed on a routine basis and provided with opportunities to be involved in the cleanup process."

In truth, the level of community involvement in the cleanup was pretty much limited to Neodesha's elected officials, who had filed suit in May 2004 demanding that BP conduct a full cleanup of the refinery waste and pay millions of dollars in damages, at a total cost to the company of up to $1 billion. Lucille Campbell, the Neodesha retiree who pushed the city to file suit after what she felt was a concerted campaign by BP to cover up the full extent of contamination, said most of the 2,800 residents of the city had lost track of the lawsuit while it slowly worked its way through the Kansas court system. The local newspaper, the *Neodesha Derrick,* provided little news about the case since, technically, as a property owner above the contaminated plume, the publisher was a plaintiff in the class action suit and eligible for a damage award. When the trial was held from late August 2007 until the beginning of January 2008, the *Derrick* asked a former state representative from Neodesha, Rochelle Chronister, to attend the trial as much as she could in nearby Erie, Kansas, and write up her observations for the paper. But after the stunning jury verdict and equally surprising reversal by Creitz, there were almost no reports about the case in the *Derrick* for two years while BP's appeal of the judge's ruling worked its way to the Supreme Court. The process was lengthy partly because it took more than half a year to prepare a transcript of the district court trial; there was also plenty of time granted for attorneys from both sides to prepare written briefs on the case, and there were dozens of motions for, first, the Kansas Court of Appeals to decide, and, finally, for the Supreme Court to handle after the appellate court kicked the case directly to it for final disposition.

Wilson County Sheriff Dan Bath was an astutely independent observer of the Neodesha proceedings, working some distance from the town in Fredonia, the county seat, but making frequent visits to the community where he grew up in the 1960s when his father worked at the Amoco refinery in Neodesha. "The pervasive smell I do remember," said Bath, fifty-three, in the fall of 2010. "People called it 'Cheesetown.'"

When Amoco closed the Neodesha refinery in 1970, Bath's father was transferred to the Texas City plant as part of the company's effort to assist the hundreds of employees put out of work by the Kansas shutdown. Both of Bath's parents died of cancer down in Texas, although both were in their eighties, and there was no way of knowing whether they had developed the disease from exposure to refinery contamination. Bath himself almost went to work at the Texas City refinery as a young man, but decided to go into law enforcement and return to his roots in Kansas.

Neodesha today is a shadow of its former days as a booming center of the oil business following the growth of the refinery built by Standard Oil Company in 1897, Bath said. "There were so many people in Neodesha who made their living from the refinery, so when it closed, half the town left," he said. "They've done an outstanding job of rebuilding, but you can't compare it to what it was." The sheriff said he believed from anecdotal evidence that cancer cases and other illnesses "have to be a little higher" in Neodesha, but there was no real certainty about the effects on people who were literally living on top of a toxic waste dump filled with deadly poisons like benzene, arsenic, and lead. "I think something needs to be done," Bath said, "but I just don't know what can be done. People who are aware of it have different degrees of concern."

The divisions in the community about the pollution were evident in the results of the city election in the spring of 2010, when the Reverend DeWayne Prosser, who had adamantly opposed the lawsuit against BP, narrowly defeated incumbent Mayor Casey Lair, who supported the litigation. "It was an issue," Prosser said. "My opponent brought up the suit." But Prosser interpreted his fifty-vote margin of victory, 310–260, as vindication for his stance against the lawsuit. "The town is still divided," he said in October of 2010, "but I think they're for the most part tired of it."

A decade earlier, Prosser had been a supporter of Lucille Campbell's efforts to push for a lawsuit against BP, feeling sympathetic to her argument that her infant daughter had died of leukemia when the family lived a block from the refinery in 1964. But over the years as a Wilson County commissioner, Prosser said he came to believe that people in the community were motivated by the desire to get a big payoff from BP, after word spread that a suburb of Kansas City, Sugar Creek, had won property damages from BP after suing over contamination from another Amoco refinery that had closed in 1982. "I think there's a whole lot of politics in this whole thing—wanting the money," Prosser said in an interview. "A billion dollars would do a lot for this whole region."

And when Lucille Campbell continued to push for a lawsuit— even bringing flyers to distribute at the Bible Baptist Church where he was pastor, Prosser alleged—he asked her to leave the congregation, on the very day her son was being ordained a minister. "She had her packets with her," Prosser insisted. "She never stopped carrying them around. She was adamant about it. But there are proper places for that, and church and her son's ordination were not the proper place."

Campbell denied that she ever brought flyers to church, realizing there were some members of the church who still relied on pensions from BP, and opposed the lawsuit. She said she was devastated when Prosser asked her to leave the congregation. "That bothered me," she said. "That was my church. I had always admired Brother Prosser, I'd prayed for him, and to think, when I realized he was totally against me—that hurt, when I had to leave my church."

Now that they were living in her grandfather's old farmhouse outside the city limits of Neodesha, Lucille Campbell and her husband Bob had no stake in the lawsuit, since only the owners of property above the plume in the city were eligible to be plaintiffs. In fact, the Campbells ran up debts from Lucille's work investigating the pollution, copying documents and printing newsletters for the community, which was part of the reason they were forced to sell her family's property to a neighboring farmer, who let the Campbells live there as long as they chose. "There were times I would think I should have kept my money for myself," Lucille said, seated in the heart of Neodesha in October 2010. "I might not have had to sell my place out there had I kept my finances better."

Lucille said she didn't even know that the death of her daughter decades earlier might have been caused by the refinery pollution until after she started her campaign to find out what was happening in the town where she had grown up. But when she dug up the hospital records of her baby Gretta's death and saw the lab report showing a white blood cell count of 44,700, a clear sign of leukemia, she knew there were serious problems in the community. "When that lab report came up 44,700, I just got kind of cold and blurted out, 'I know why she died.' We were living right there for a year and a half," in a house just a block from the refinery when Gretta was born and died after less than four months, she said. Another daughter who also lived in the house was later diagnosed with lupus, she added.

Lucille wasn't sure what should be done about the contamination in Neodesha, but she felt something needed to be done to reduce people's exposure to it. "They could move people out to the country," she said. "There have been other towns that have done that."

Lucille Campbell was unable to attend when the Kansas Supreme Court heard oral arguments on *Neodesha v. BP* on October 27. Her husband was dying, at home under hospice care, and in great pain from lung cancer and a variety of other health problems, partly related to a bad smoking habit earlier in his nearly eighty years on the Great Plains.

The arguments over the case in the spacious Kansas Judicial Center in Topeka would only last for an hour, which seemed like a blip in the more than six years since the lawsuit was first filed. Nevertheless, more than two dozen Neodesha residents and officials, plus an entire business law class from Neodesha High School, made the two-hour drive to the state capital to attend the hearing, which would help determine the fate of one of the nation's original oil towns, where the first commercial well field west of the Mississippi River had been tapped in 1892.

BP's attorney, Rick Godfrey of Kirkland & Ellis in Chicago, made the initial presentation, and openly expressed incredulity at Judge Creitz's decision to throw out the jury's unanimous verdict that BP had fulfilled its obligations to the city and was no longer liable for further cleanup in Neodesha. Judge Creitz seemed to ignore the fact that the Kansas Department of Health and Environment had approved every aspect of cleanup work the company had done in Neodesha, Godfrey said. It was a "state-supervised, state-requested, state-approved, state-sanctioned remediation and cleanup," yet Creitz still ruled that BP was engaged in "abnormally dangerous activity" by failing to remove all contaminants from the refinery site.

Godfrey asked the court to reject Creitz's ruling calling for a new trial on how much more BP should spend to clean up the town, and let the jury's finding that the company was not liable for further remediation stand.

The attorney for Neodesha, Daniel Young, argued that Creitz was correct in dismissing the jury's findings because BP had publicly acknowledged it was liable for the pollution and had promised to clean it up. Instead, the company did little more than pump out some of the contaminated groundwater, leaving the bulk of the waste to take care of itself through a process the company called "natural attenuation," Young said. "They were aware there was oil under the refinery because someone dug a hole and it caught fire," he said. "In 2000 they dug a trench and it filled with oil. When there's oil in the ground, monitoring and natural attenuation won't work. But it is cheaper for the polluter to let nature take its course."

The court needed to uphold Creitz's ruling to maintain the integrity of state environmental law, Young argued. "The message to the polluter is if you say you're going to clean it up, and you accept responsibility for it, you have to proceed honestly, you have to proceed diligently, you have to do the job right," he said. "Just saying you're going to remediate and hiring a public relations firm is a lot cheaper than actually cleaning it up. That's what they did."

The Supreme Court justices listened to the arguments for the allotted hour, asked some questions on fine points of the law surrounding the case, and moved on to the next hearing without any indication of how they might rule. An opinion was expected at any time, but had not been issued by the end of 2010.

In the late-evening hours of Sunday, November 14, Bob Campbell died with his wife Lucille at his bedside. "I put on some gospel music and told him, 'Go on to your mama, it's okay,'" Lucille said the next morning.

Now Lucille Campbell was looking ahead to the long, cold Kansas winter, alone in her grandparents' house that she no longer owned, rejected by the church that she had loved, and still uncertain whether BP would finally be held accountable for the vast plume of petroleum wastes that she believed was killing her hometown.

NOTES

chapter two

1. Ronald W. Ferrier, *The History of the British Petroleum Company, Vol. 1: The Developing Years, 1901–1932* (Cambridge: Cambridge University Press, 1982).

2. Berry Ritchie, *Portrait in Oil: An Illustrated History of BP* (London: Brittanic House, 1995), 32–33.

3. Daniel Yergin, *The Prize: The Epic Quest for Oil, Money & Power* (New York: Free Press, 1991), 457.

4. Yergin, *The Prize*, 460.

5. Kermit Roosevelt, *Countercoup: The Struggle for Control of Iran* (New York: McGraw-Hill, 1979), 156–57.

6. Ritchie, *Portrait in Oil*, 79.

7. Ritchie, *Portrait in Oil*, 103–105.

8. John Browne, *Beyond Business: An Inspirational Memoir from a Visionary Leader* (London: Wiedenfeld & Nicolson, 2010), 50–54.

9. *Multinational Monitor*, "BP: A Legacy of Aparthied, Pollution and Exploitation," November 1992.

10. Notes from interviews provided by Ernest A. Lowe, coauthor with Robert J. Harris of "Taking Climate Change Seriously: British Petroleum's Business Strategy," *Corporate Environmental Strategy* (Winter 1998).

11. Browne, *Beyond Business,* 67.

12. Darcy Frey, "How Green is BP?" *The New York Times Magazine,* December 8, 2002.

chapter three

1. *Houston Chronicle,* "Unit at Refinery Has Troubled Past," April 11, 2005.

2. *Houston Chronicle,* "Horror in Trailer Bonds BP Survivors," November 27, 2005.

3. Mimi Swartz, "Eva vs. Goliath," *Texas Monthly,* July 2007.

4. *Houston Chronicle,* "BP Leads Nation in Refinery Fatalities," May 15, 2005.

5. *Houston Chronicle,* "BP Says Its Initial Findings Misstated," May 25, 2005.

6. *Houston Chronicle,* "Federal Investigators Begin Probe of Latest BP Explosion," July 29, 2005.

7. *Houston Chronicle,* "Latest BP Blast Site Had Record of Inadequate Maintenance," August 7, 2005.

8. Mimi Swartz, "Eva vs. Goliath."

9. *Houston Chronicle,* "Horror in Trailer Bonds BP Survivors."

10. *Ladies' Home Journal,* "Rebel with a Cause," September 2007.

11. Mimi Swartz, "Eva vs. Goliath."

12. *Ladies' Home Journal,* "Rebel with a Cause."

13. *Houston Chronicle,* "Lawsuit Yields Millions, but Woman 'Hates BP'," April 24, 2007.

chapter four

1. John Browne, *Beyond Business: An Inspirational Memoir from a Visionary Leader* (London: Wiedenfeld & Nicolson, 2010), 33.

2. *Anchorage Daily News,* "Corrosion Infested Prudhoe Pipeline," October 31, 2006.

chapter five

1. Howard Gardner, *Changing Minds: The Art and Science of Changing Our Own and Other People's Minds* (Boston: Harvard Business School Press, 2004), 143.

2. *Evening Standard*, "City Pension Fund Sues 'Negligent' BP," October 9, 2006.

3. *The Wall Street Journal*, "U.S. Accuses BP of Manipulating Price of Propane," June 29, 2006.

4. *Chicago Tribune*, "Troubles Run Deep on Gulf Oil Platform," May 28, 2007.

5. *The New York Times*, "Drilling Deep in the Gulf of Mexico," November 8, 2006.

6. *Houston Chronicle*, "Federal Investigator Recites Causes of BP Blast," March 20, 2007.

7. *Chicago Tribune*, "BP Dumps Mercury in Lake," July 27, 2007.

8. Associated Press, "BP Plant Explosion Lawsuit Ends in Settlement," September 18, 2007.

9. *Newsweek*, "Slick Operator," May 7, 2010.

chapter seven

1. *Houston Chronicle*, "Wife of Worker Killed This Year at BP Plant Says Husband Was Thinking Ahead to Finishing Career," January 31, 2008.

2. Dow Jones Newswire, "Internal BP List Suggests Spill, Leaks Up Since CEO Took Over," February 4, 2008.

3. *The Sunday Times*, "Tony Hayward Makes His Mark on BP," November 1, 2009.

4. *Houston Chronicle*, "BP Blast Survivors Line Up to Speak," January 30, 2008.

5. Reuters, "Appeals Court Says BP Blast Deal Violates U.S. Law," May 7, 2008.

6. Reuters, "U.S. Judge Approves BP Guilty Plea in 2005 Explosion," March 12, 2009.

7. *The Independent*, "BP Chairman Is Like Goebbels, Says Russian Oligarch," June 17, 2008.

8. Reuters, "Third of BP Investors Oppose Pay Plan, RBS Looms," April 16, 2009.

9. *The Independent*, "Big Guns Fail to Take Bait in BP Chairman Hunt," April 12, 2009.

10. *The Times of London*, "BP Chief Signals Renewables Profit Drive," February 28, 2008.

11. *Houston Chronicle*, "BP Chief Spells Out Oil Giant's Strategy," February 27, 2008.

12. Associated Press, "BP Chief: Industry, Government Must Cooperate," February 10, 2009.

13. Green Tech (CNET), "BP Scientist: To Cut Oil Use, Make Carbon Expensive," September 22, 2008.

14. *In These Times*, "Oil Giant BP's Role in 'Biggest Environmental Crisis'," March 31, 2008.

15. *The Guardian*, "Tony Hayward: BP's Straight-talking Chief on Evolution not Revolution," February 4, 2010.

16. *Corporate Social Responsibility News*, "Investors Decry BP's Entry Into Tar Sands; Statement to Be Submitted at BP Annual Meeting Today in London," April 16, 2008.

17. *The Telegraph*, "BP Still a Good Friend to the Environment, Says Lord Browne," February 14, 2010.

18. *Bloomberg News*, "BP Makes 'Giant' Oil Discovery in Gulf of Mexico," September 2, 2009.

19. *The Wall Street Journal*, "As CEO Hayward Remade BP, Safety, Cost Drives Clashed," June 29, 2010.

chapter eight

1. House Energy and Commerce Committee reports, 2010.

2. Hearings of the Joint Commission of the U.S. Coast Guard and the Bureau of Ocean Energy Management Investigating the Gulf of Mexico Spill.

3. *The Wall Street Journal*, "On Doomed Rig's Last Day, a Divisive Change of Plan," August 26, 2010.

4. *The Wall Street Journal*, "On Doomed Rig's Last Day."

5. Hearings of the Joint Commission.

6. CBS, *60 Minutes*, "Blowout: The Deepwater Horizon Disaster," May 16, 2010.

7. *The Wall Street Journal*, "On Doomed Rig's Last Day."

8. *The Wall Street Journal,* "BP Decisions Set Stage for Disaster," May 27, 2010.

9. *The Wall Street Journal,* "There Was 'Nobody in Charge,'" May 27, 2010.

10. CBS, *60 Minutes,* May 16, 2010.

11. *The Wall Street Journal,* "There Was 'Nobody in Charge.'"

12. Ibid.

13. *The New York Times,* "Regulators Failed to Address Risks in Oil Rig Fail-Safe Device," June 20, 2010.

14. *Los Angeles Times,* "Rig Mechanic Says BP Was in a Rush Despite Problems," May 26, 2010.

15. *The Washington Post,* "Oil Riggers on Ship That Exploded in Gulf of Mexico Describe Fateful Night," May 7, 2010.

16. *The Wall Street Journal,* "There Was 'Nobody in Charge.'"

17. *The New York Times,* "Inspector General's Inquiry Faults Regulators," May 24, 2010.

18. *Los Angeles Times,* "Federal Regulators Haven't Kept Up with Oil Drilling Expansion," May 8, 2010.

19. *McClatchy Newspapers,* "U.S. Agency Lets Oil Industry Write Offshore Drilling Rules," May 10, 2010.

20. *The Wall Street Journal,* "Rig Owner Had Rising Tally of Accidents," May 10, 2010.

21. *The Washington Post,* "Pressure Grows for Action by BP," May 1, 2010.

22. *The Wall Street Journal,* "BP Revised Permits Before Blast," May 30, 2010.

23. *The New York Times,* "Documents Show Early Worries About Safety of Rig," May 29, 2010.

24. *The Wall Street Journal,* "BP Relied on Cheaper Wells," June 19, 2010.

25. House Energy and Commerce Committee reports.

26. *The Times-Picayune,* "Costly, Time-consuming Test of Cement Linings in Deepwater Horizon Rig Was Omitted, Spokesman Says," May 19, 2010.

27. *Bloomberg News,* "BP Says Transocean, Halliburton Share Blame in Spill," September 8, 2010.

28. *The New York Times,* "Regulators Failed to Address Risks in Oil Rig Fail-Safe Device."

chapter nine

1. *The New York Times,* "Search Continues After Oil Rig Blast," April 21, 2010.

2. Center for Public Integrity, "Coast Guard Logs Reveal Early Spill Estimate of 8,000 Barrels a Day," June 3, 2010.

3. *Los Angeles Times,* "BP's Containment Problem Is Unprecedented," April 30, 2010.

4. *The New York Times,* "Regulators Failed to Address Risks in Oil Rig Fail-Safe Device," June 20, 2010.

5. *The Times-Picayune,* "Five Times as Much Oil Spewing in Gulf of Mexico Oil Spill as First Thought," April 28, 2010.

6. Ibid.

7. *The New York Times,* "Oil Spill's Blow to BP's Image May Eclipse Costs," April 29, 2010.

8. Ibid.

9. *Los Angeles Times,* "For BP, Oil Spill Is a Public Relations Catastrophe," April 30, 2010.

10. Ibid.

11. *Financial Times,* "BP Faces Revolt Over Clean-up Jobs," July 2, 2010.

12. *PBS NewsHour with Jim Lehrer,* June 16, 2010.

13. Associated Press, "BP Asks for Defense Dept. Help in Gulf Oil Spill," April 29, 2010.

14. Associated Press, "BP Goes on the Defensive as Spill Spreads," May 3, 2010.

15. Associated Press, "BP Asks for Defense Dept. Help in Gulf Oil Spill."

16. *The Times-Picayune,* "Plaquemines Parish President Billy Nungesser Becomes the Face of Oil Spill Frustration," June 6, 2010.

17. *The Times-Picayune,* "Long-term Impact of Gulf of Mexico Oil Spill Remains Unclear," April 30, 2010.

18. *The Times-Picayune,* "Gulf Spill Is Really a River of Oil, Environmentalists Say," April 29, 2010.

19. *The Washington Post,* "Officials' Forecast Grim About Massive Oil Spill as Obama Tours Part of the Gulf Coast," May 3, 2010.

20. *The Wall Street Journal,* "BP's Worsening Spill Crisis Undermines CEO's Reforms," May 3, 2010.

21. *Mobile Press-Register,* "BP Told to Stop Circulating Settlement Agreements with Coastal Alabamians," May 2, 2010.

22. CBS, "BP Told to Stop Distributing Oil Spill Settlement Agreements," May 3, 2010.

23. Reuters, "U.S. Regulators Sued Over BP's Atlantis Platform," May 17, 2010.

24. *The Wall Street Journal,* "BP's Preparedness for Major Crisis Is Questioned," May 10, 2010.

25. *The Washington Post,* "Pressure Grows for Action by BP," May 1, 2010.

26. *McClatchy Newspapers,* "Top U.S. Offshore Drilling Official Abruptly Retires," May 18, 2010.

27. *The Guardian,* "BP Boss Admits Job on the Line over Gulf Oil Spill," May 14, 2010.

28. *Sky News,* "BP Chief Predicts 'Very Modest' Oil Spill Impact," May 18, 2010.

29. *The New York Times,* "Giant Plumes of Oil Found Forming Under Gulf of Mexico," May 15, 2010.

30. *McClatchy Newspapers,* "Low Oil Spill Estimate Could Save BP Millions in Court, Experts Say," May 20, 2010.

31. *The Washington Post,* "Challenge of Cleaning Up Gulf of Mexico Oil Spill 'Unprecedented' at Such Depths," May 15, 2010.

32. Ibid.

33. *PBS NewsHour with Jim Lehrer,* May 25, 2010

34. *Rolling Stone,* "The Spill, the Scandal and the President," June 8, 2010.

35. Associated Press, "After 'Top Kill' Failure, BP Works to Stem Gulf Leak," May 31, 2010.

36. Ibid.

37. *Los Angeles Times,* "Kevin Costner May Hold Key to Oil Spill Cleanup," May 21, 2010.

38. New York *Daily News,* "James Cameron Says BP Turned Away His Offer to Help with Gulf Oil Spill," June 3, 2010.

39. *USA Today,* "After BP Oil Spill, Thousands of Ideas Poured in for Cleanup," November 15, 2010.

40. ABC, "BP Oil Spill: Residents of Marrero, La., Share Their Frustration and Fears," May 24, 2010.

41. *Fox News Sunday With Chris Wallace,* May 23, 2010.

42. *The Washington Post,* "Sarah Palin's Fact-free Commentary on Paul and BP," May 26, 2010.

43. CNN, *State of the Union with Candy Crowley,* May 30, 2010.

44. NBC, *Meet the Press,* May 30, 2010.

45. ProPublica, "EPA Officials Weigh Sanctions Against BP's U.S. Operations," May 21, 2010.

46. CNN, "'Top Kill' Fails, BP Moves On 'To Next Option,'" May 29, 2010.

47. *The New York Times,* "Efforts to Repel Oil Spill Are Described as Chaotic," June 14, 2010.

48. CBS, *The Early Show,* May 28, 2010.

49. Reuters, "Inside BP's War Room," June 1, 2010.

50. *Bloomberg News,* "BP Credibility Questions Grow as U.S. Lawmakers Press Inquiry," May 31, 2010.

51. Associated Press, "BP's Top Kill Effort Fails to Plug Gulf Oil Leak," May 29, 2010.

52. *Los Angeles Times,* "Gloom Grows as BP's 'Top Kill' Effort Fails," May 30, 2010.

53. *McClatchy Newspapers,* "As Oil Keeps Flowing from Gulf of Mexico Leak, Criticism Is Growing," May 29, 2010.

54. *Fox News Sunday with Chris Wallace,* May 31, 2010

55. *New York Daily News,* "BP's CEO Tony Hayward: The Most Hated—and Most Clueless—Man in America," June 2, 2010.

56. NBC, *Today,* June 8, 2010.

57. *Times of London,* "That's Enough 'Kicking Ass', Mr. President," June 11, 2010.

58. *The New York Times,* "BP Chief Draws Outrage for Attending Yacht Race," June 19, 2010.

59. ABC, *This Week,* June 20, 2010.

60. Fox News, June 19, 2010.

61. *Bloomberg News,* "BP Oil Disaster Costs U.S. State Pensions $1.4 Billion in Value," June 22, 2010.

62. *The New York Times,* "As BP Staggers, Pension Funds Skid," June 18, 2010.

63. Reuters, "Should BP Nuke Its Leaking Well?" July 2, 2010.

64. *The Wall Street Journal,* "BP Shares Rise as Oil Leak Is Halted," July 15, 2010.

65. *Christian Science Monitor,* "Joe the Plumber (Not That One) Says He Helped Stop Gulf Oil Spill Leak," July 17, 2010.

66. ABC News, "BP Buys 'Oil' Search Terms to Redirect Users to Official Company Website," June 5, 2010.

chapter ten

1. Hearings before Joint Commission of Coast Guard and Bureau of Ocean Energy Management, July 22, 2010.

2. *Natchez Democrat,* "Family Left with Only Memories," May 2, 2010.

3. *Matagorda Advocate,* Anderson Obit, May 20, 2010.

4. *Los Angeles Times,* "Town Grieves for Oil Rig Worker Lost in Explosion," May 8, 2010.

5. *Esquire,* "Eleven Lives," August 16, 2010.

6. Associated Press, "At the Heart of Oil Spill, 11 Grieving Families," May 5, 2010.

7. *Bloomberg News,* "BP Pressured Rig Worker to Hurry Before Disaster, Father Says," May 27, 2010.

8. *AolNews,* "Relatives Fear the Dead Oil Rig Workers Are Forgotten," May 23, 2010.

9. Associated Press, "At the Heart of Oil Spill, 11 Grieving Families."

10. *AolNews,* "Relatives Fear the Dead Oil Rig Workers Are Forgotten."

11. ABC News, *Good Morning America,* June 10, 2010.

12. ABC News, White House blog, June 10, 2010.

13. *The Times-Picayune,* "Father of Deepwater Horizon Victim Reportedly Joined Call for Ouster of BP CEO Tony Hayward," June 2, 2010.

14. Associated Press, "Suffering Continues for Oil Disaster Survivors," July 20, 2010.

15. *The New York Times,* "Workers on Doomed Rig Voiced Concern About Safety," July 21, 2010.

16. *Bloomberg News*, "BP Well Boss Won't Shed Light on Cause of Oil Spill," July 19, 2010.

17. Hearings before Joint Commission of Coast Guard and Bureau of Ocean Energy Management.

18. Environment America, "Too Much At Stake: Don't Gamble With Our Coasts," November 16, 2010.

19. NOAA, "Gulf of Mexico Oil Spill, Economic Impacts to Fisheries and Coastal Habitat," April 2010.

20. National Marine Fisheries Service, Annual Commercial Landing Statistics.

21. National Marine Fisheries Service, Gulf of Mexico summary.

22. Natural Resources Defense Council, "A Fisherman Pirate Fights for His Life in the Bayou," July 28, 2010.

23. *Los Angeles Times*, "Gulf Fishermen Aren't Ready to Ditch BP," August 6, 2010.

24. *Huffington Post,* "A Forgotten Fisherman Fights to Survive the BP Oil Disaster," November 4, 2010.

25. *Louisiana Bayoukeeper,* "A Fisherman's Heartbreak: Louisiana's Coming 'Summer of Tears,'" May 25, 2010.

26. Associated Press, "Huge Estuary Now Ground Zero in Oil Spill," June 14, 2010.

27. Associated Press, "Head of BP Fund Hears about Seafood Industry Loss," July 24, 2010.

28. Associated Press, "Spill Puts Obama's Oil Fund Chief on Hostile Turf," July 28, 2010.

29. *The New York Times,* "Spill Takes Toll on Gulf Workers' Psyches," June 16, 2010.

30. Ibid.

31. CNN, "Vietnamese Fishermen in Gulf Fight to Not Get Lost in Translation," June 24, 2010.

32. WGNO, "Louisiana Fishermen Contemplating Suicide, Need Mental Health Services," May 21, 2010.

33. Associated Press, "Vietnamese 'Lost' as Gulf Oil Spill Hits Community Hard," July 11, 2010.

34. Associated Press, "Spill Reinforces Oil Bad Will for American Indians," May 18, 2010.

35. *Epoch Times*, "BP Oil Spill Taking Toll on Louisiana Indian Tribe," October 31, 2010.

36. KTUU, "Louisianans Visit Alaska for Help Coping with Gulf Oil Spill," August 2, 2010.

37. *The New York Times*, "Fishermen Fear Disruption of Their Way of Life," May 29, 2010.

38. *Time*, "Is It Twilight for Louisiana's Black Oystermen?" June 8, 2010.

39. *The Times-Picayune*, "Oyster Harvester Trying to Find His Footing 5 Months After Gulf Oil Spill," October 6, 2010.

40. *The Guardian*. "BP Spill: White House Says Oil Has Gone, But Gulf's Fishermen Are Not So Sure," September 6, 2010.

41. Ibid.

42. CNN, "BP Cuts Checks for Workers' Lost Wages," June 1, 2010.

43. *The Guardian*. "BP Spill: White House Says Oil Has Gone, But Gulf's Fishermen Are Not So Sure."

44. Reuters, "Gulf Fishermen: Oil Tainted Our Waters, Our Trust," August 12, 2010.

45. *The Times-Picayune*, "Scientists Wary of BP Oil Spill's Long-term Effects on Species," November 10, 2010.

46. Reuters, "Gulf Fishermen: Oil Tainted Our Waters, Our Trust."

47. *USA Today*, "Gulf Seafood's Falling Supply, Demand Disrupt Industry," October 8, 2010.

48. *Bloomberg News*, "Assessing the After-Effects of the BP Oil Spill," September 14, 2010.

49. *Daily Comet*, "Seafood Officials Say They Need More BP Aid," September 11, 2010.

50. *The Wall Street Journal*, "Oysters Lose Their Allure," November 6, 2010.

51. *Bloomberg News*, "Gulf Coast Oil Spill 'Terrifying' for Developers from Louisiana to Florida," May 4, 2010.

52. *The Wall Street Journal*, "Gulf Business Owners Critical of BP Claim Process," June 10, 2010.

53. *Bloomberg News,* "Gulf Coast Oil Spill 'Terrifying' for Developers from Louisiana to Florida."

54. *Los Angeles Times,* "BP Claims Absolute Responsibility for Oil Cleanup," May 4, 2010.

55. *USA Today,* "Travel Industry Group Wants BP to Pay $500 Million," July 22, 2010.

56. CNS News, "No Evidence Gulf Oil Spill Killed Fish, Says NOAA," October 20, 2010.

57. *The Wall Street Journal,* "Gulf Business Owners Critical of BP Claim Process."

58. Associated Press, "Head of BP Fund Hears About Seafood Industry Loss," July 24, 2010.

59. *The Times-Picayune,* "Churches, Nonprofits Fight for Survival in Face of Gulf of Mexico Oil Spill," July 3, 2010.

60. *Washington Independent,* "Gulf Coast Residents in Financial Dire Straits, Waiting for BP Claims," September 13, 2010.

61. National Public Radio, "Gulf Coast Claimants Irked by BP's Delays," August 9, 2010.

62. *Los Angeles Times,* "Gulf Spill Claims: Take Money Now, Or Chances Later?," August 24, 2010.

63. CNN, "New Orleans Chef Susan Spicer Suing BP," June 29, 2010.

64. *Birmingham Business Journal,* "Fish Market Owner Suing BP," August 30, 2010.

65. *Inter Press Service,* "Illness Plagues Gulf Residents in BP's Aftermath," November 15, 2010.

66. *Louisiana Daily Planet,* "Illness from Spill Unknown Still," September 19, 2010.

67. *Inter Press Service,* "Broad Coalition Rallies for BP Accountability," November 1, 2010.

68. *Fortune,* "Why a Flounder Gigger's Suit Against BP Is One to Watch," July 30, 2010.

69. *The Washington Post,* "Officials Ask BP to Protect Health of Workers Cleaning Up Oil Spill," May 28, 2010.

70. *Los Angeles Times,* "Oil Cleanup Workers Report Illness," May 26, 2010.

71. *Los Angeles Times,* "Gulf Oil Spill: Human Health Effects Debated," June 4, 2010.

72. *Los Angeles Times,* "Oil Cleanup Workers Report Illness."

73. *Bloomberg News,* "Gaps in Health Data, Suspicions About BP Worry U.S. Panelists at Hearing," June 22, 2010.

74. *Greenwire,* "Locals Track Gulf Oil Spill's Health Impacts, Paving Way for Federal Study," August 19, 2010.

75. *Our Sunday Visitor,* "Church Tends to Oil Spill Victims' Physical and Spiritual Needs," June 3, 2010.

76. *Mobile Press-Register,* "After Oil Spill, Depression and Stress Levels Rise in Coastal Alabama," September 27, 2010.

77. *Our Sunday Visitor,* "Church Tends to Oil Spill Victims' Physical and Spiritual Needs."

78. *Los Angeles Times,* "Oil Spill Stress Starts to Weigh on Gulf Residents," June 20, 2010.

79. *Our Sunday Visitor,* "Church Tends to Oil Spill Victims' Physical and Spiritual Needs."

80. National Public Radio, "Gulf Residents Cope With Mental Health Issues," July 8, 2010.

81. *Los Angeles Times,* "Suicide Is Called Another Casualty of BP Oil Spill," June 25, 2010.

chapter eleven

1. *Time,* "The BP Spill: Has the Damage Been Exaggerated?" July 29, 2010.

2. *Nature,* "Deepwater Horizon: After the Oil," September 1, 2010.

3. CNS News, "No Evidence Gulf Oil Spill Killed Fish, Says NOAA," October 20, 2010.

4. Ibid.

5. WWL-TV, "Scientists Question Thoroughness of Gulf Seafood Testing," September 20, 2010.

6. *McClatchy Newspapers,* "Gulf Oil Spill Still Poses Health Threats: Study," August 16, 2010.

7. *The News Star,* "Barham: We Can't Trust BP," August 26, 2010.

8. *Mobile Press-Register,* "Baby Fish Show Up in Big Numbers Despite Gulf of Mexico Oil Spill," September 26, 2010.

9. CNS News, "No Evidence Gulf Oil Spill Killed Fish, Says NOAA."

10. *Huffington Post,* "The Crime of the Century: What BP and the US Government Don't Want You to Know, Part I," August 4, 2010.

11. *Audubon,* "How Many Birds Died in the BP Oil Spill?" October 28, 2010.

12. *Rolling Stone,* "The Poisoning," July 21, 2010.

13. National Public Radio, *All Things Considered,* October 15, 2010.

14. Associated Press, "Scientists Say Gulf Spill Altering Food Web," July 14, 2010.

15. Associated Press, "Crabs Provide Evidence Oil Tainting Gulf Food Web," August 9, 2010.

16. *Science Times,* "BP Gusher Left Deep Sea Toxic for a Time, Study Finds," November 1, 2010.

17. *Nature,* "Oil Spill's Toxic Trade-off," November 2010.

18. *Los Angeles Times,* "BP Jams Gulf Well with Drilling Mud," August 5, 2010.

19. *The Independent,* "Oil Spill Creates Huge Undersea 'Dead Zones,'" May 30, 2010.

20. *The Washington Post,* "Scientists Report Undersea Oil Plume Stretching 21 Miles from BP Spill Site," August 19, 2010.

21. *The New York Times, Greenwire,* "Natural Gas From Ruptured Gulf Well Remained Trapped in Deep Waters," September 17, 2010.

22. Reuters, "U.S. Govt: No Sign of Undersea Plume from BP Spill," September 21, 2010.

23. National Public Radio, *All Things Considered,* September 10, 2010.

24. ABC News, "Oil from the BP Spill Found at Bottom of Gulf," September 12, 2010.

25. *USA Today,* "Research Teams Find Oil on Bottom of Gulf," October 25, 2010.

26. *The Times-Picayune,* "Federal Leaders of Gulf of Mexico Oil Spill Response Report Only a Few Lingering Trouble Spots," October 18, 2010.

27. *USA Today,* "Research Teams Find Oil on Bottom of Gulf."

28. Associated Press, "BP Deep-cleaning Gulf Beaches Amid New Worries," November 17, 2010.

29. *Nature,* "Freedom of Spill Research Threatened," July 28, 2010.

30. National Public Radio, *Weekend Edition,* "By Hiring Gulf Scientists, BP May Be Buying Silence," July 31, 2010.

chapter twelve

1. ProPublica, "BP Texas Refinery Had Huge Toxic Release Just Before Gulf Blowout," July 2, 2010.

2. *The New York Times,* "With Neighbors Unaware, Toxic Spill at a BP Plant," August 29, 2010.

3. ProPublica, "BP Texas Refinery Had Huge Toxic Release Just Before Gulf Blowout."

4. *Galveston County Daily News,* "143 File Lawsuit Against BP," April 9, 2009.

5. *Houston Chronicle,* "State Sues BP, Alleging Texas City Pollution Violations," June 5, 2009.

6. *Courthouse News Service,* "Oil Giant BP Sued by Shareholders for Engaging in Extreme Environmental-Law Violations, Which Drove Down Stock Value," August 6, 2009.

7. Reuters, "BP Ordered to Pay $100 million in Chemical Case," December 19, 2009.

8. *Southeast Texas Record,* "Workers Seek $500 Million Over Benzene Vapor Release at Texas City Refinery," January 20, 2010.

9. *Houston Chronicle,* "Thousands Sign On for $10 billion BP Suit," August 6, 2010.

10. *Galveston County Daily News,* "OSHA Fine Raises Fears of BP Shutdown," November 8, 2009.

11. Center for Public Integrity, "Renegade Refiner: OSHA Says BP Has 'Systemic Safety Problem,'" May 16, 2010.

12. Reuters, "Alaska Criticizes BP over Natgas Rupture Follow-up," February 25, 2009.

13. *Anchorage Daily News,* "Judge Could Revoke BP's Probation in North Slope Spills," November 19, 2010.

14. ProPublica, "With All Eyes on the Gulf, BP Alaska Facilities Are Still at Risk," November 2, 2010.

15. Truthout, "New Documents, Employees Reveal BP's Alaska Oilfield Plagued by Major Safety Issues," June 15, 2010.

16. *Business Week,* "The Oil Spill: Will BP Face Criminal Charges?" July 1, 2010.

17. Associated Press, "RICO Law Made to Combat Mafia Used in BP Lawsuits," July 20, 2010.

18. *Financial Times,* "BP and Halliburton Face Bigger Claims," October 29, 2010.

19. *The Washington Post,* "Lawyers Lining Up for Class-action Suits Over Oil Spill," May 17, 2010.

20. Reuters, "BP Approaches Funds to Fend Off Takeover Bids: Source," July 6, 2010.

21. Reuters, "New York Fed Probes Wall Street Exposure to BP, Say Sources," June 28, 2010.

22. Reuters, "British PM Fears BP's 'Destruction', Stock Plunges," June 25, 2010.

23. *Bloomberg News,* "BP's Dudley Embraces Deepwater Risk in U.S., Brazil After Spill," November 3, 2010.

24. *Mobile Press-Register,* "BP Will Not Use Well Responsible for Oil Spill, Says Incoming CEO," August 29, 2010.

25. Cambridge Energy Research Associates, "Growth in the Canadian Oil Sands: Finding the New Balance," 2009.

26. *The Guardian,* "Gulf Oil Spill: Will Deepwater Sink the 101-Year-Old BP?" June 2, 2010.

SOURCES

Alaska Oil Spill Commission final report, 1990.

BP press releases from http://www.bp.com.

Brent Coon & Associates, documents obtained in BP litigation.

U.S. Occupational Safety and Health Administration reports and news releases.

The report of the BP U.S. Refineries Independent Safety Review Panel.

Testimony by Charles Hamel before the House Interior and Insular Affairs Committee, November 1991.

Report of the House Interior and Insular Affairs Committee on Alyeska Pipeline Service Company Covert Operation, July 1992.

Congressional hearings on Alyeska pipeline problems, July and November 1993.

Reports by Richard A. Fineberg at FinebergResearch.com.

Coffman Engineers, technical analysis of BP Exploration (Alaska) Inc. corrosion monitoring, June 2002.

Vinson & Elkins report for BP Exploration (Alaska) on allegations of workplace harassment, October 2004.

The Telos Group, BP Texas City Site Report of Findings, January 2005.

House Energy and Commerce Subcommittee on Oversight and Investigations hearings on BP pipeline problems at Prudhoe Bay, September 2006 and May 2007.

U.S. Chemical Safety and Hazard Investigation Board report on BP
Texas City refinery explosion and fire, March 2007.

Press releases and statements by members of Congress from public
Web sites.

Neodesha v. BP trial transcript and related documents, Wilson
County, Kansas, District Court and Kansas Supreme Court.

House Energy and Commerce Committee documents and hearings on
BP oil spill in the Gulf of Mexico, 2010.

Deepwater Horizon Joint Investigation by the Coast Guard and Bureau
of Ocean Energy Management, reports and hearings, 2010.

National Commission on the BP Deepwater Horizon Oil Spill and
Offshore Drilling, reports and hearings, 2010.

Testimony before the House Energy and Commerce Subcommittee on
Oversight and Investigations field hearing on the local impacts
of the BP oil spill, Chalmette, Louisiana, June 2010.

Press releases on Gulf of Mexico studies from Penn State University,
Project Seahorse, and the Dauphin Island Sea Lab.

White House briefing transcripts.

Press releases and statements from the Ocean Conservancy, the Nature
Conservancy, and the U.S. Environmental Protection Agency.

Senate Health, Education, Labor, and Pensions Committee, hearing of
the Subcommittee on Employment and Workplace Safety,
June 2010.

BP Deepwater Horizon Accident Investigation Report, September 2010.

Transocean and Halliburton press releases on BP report.

Alliance for Justice report, "Judicial Ties to the Oil Industry & the Duty
to Recuse," October 2010.

BP CEO Robert Dudley speech to Confederation of British Industry,
October 2010.

National Academy of Engineering and National Research Council
interim report on the causes of the Deepwater disaster,
November 2010.

INDEX